Observations and Theory of Short GRBs at the Dawn of the Gravitational Wave Era

Observations and Theory of Short GRBs at the Dawn of the Gravitational Wave Era

Special Issue Editors

Giulia Stratta
Andrea Rossi
Maria Giovanna Dainotti

MDPI • Basel • Beijing • Wuhan • Barcelona • Belgrade

Special Issue Editors

Giulia Stratta
Urbino University
Italy

Andrea Rossi
INAF-OAS Bologna
Italy

Maria Giovanna Dainotti
Jagiellonian University
Poland

Editorial Office
MDPI
St. Alban-Anlage 66
4052 Basel, Switzerland

This is a reprint of articles from the Special Issue published online in the open access journal *Galaxies* (ISSN 2075-4434) from 2018 to 2019 (available at: https://www.mdpi.com/journal/galaxies/special_issues/GammaRayBursts).

For citation purposes, cite each article independently as indicated on the article page online and as indicated below:

LastName, A.A.; LastName, B.B.; LastName, C.C. Article Title. *Journal Name* **Year**, *Article Number*, Page Range.

ISBN 978-3-03921-291-0 (Pbk)
ISBN 978-3-03921-292-7 (PDF)

Cover image courtesy of NSF/LIGO/Sonoma State University/A. Simonnet.

© 2019 by the authors. Articles in this book are Open Access and distributed under the Creative Commons Attribution (CC BY) license, which allows users to download, copy and build upon published articles, as long as the author and publisher are properly credited, which ensures maximum dissemination and a wider impact of our publications.

The book as a whole is distributed by MDPI under the terms and conditions of the Creative Commons license CC BY-NC-ND.

Contents

About the Special Issue Editors vii

Preface to "Observations and Theory of Short GRBs at the Dawn of the Gravitational Wave Era" ix

Antonios Nathanail
Binary Neutron Star and Short Gamma-Ray Burst Simulations in Light of GW170817
Reprinted from: *Galaxies* **2018**, *6*, 119, doi:10.3390/galaxies6040119 1

Soheb Mandhai, Nial Tanvir, Gavin Lamb, Andrew Levan and David Tsang
The Rate of Short-Duration Gamma-Ray Bursts in the Local Universe
Reprinted from: *Galaxies* **2018**, *6*, 130, doi:10.3390/galaxies6040130 33

Massimiliano De Pasquale
The Host Galaxies of Short GRBs as Probes of Their Progenitor Properties
Reprinted from: *Galaxies* **2019**, *7*, 30, doi:10.3390/galaxies7010030 49

Takanori Sakamoto, Yuuki Yoshida and Motoko Serino
Investigation of Similarity in the Spectra between Short- and Long-Duration Gamma-ray Bursts
Reprinted from: *Galaxies* **2018**, *6*, 106, doi:10.3390/galaxies6040106 57

Nicole Lloyd-Ronning
Reverse Shock Emission from Short GRBs
Reprinted from: *Galaxies* **2018**, *6*, 103, doi:10.3390/galaxies6040103 67

About the Special Issue Editors

Giulia Stratta Expert of gamma-ray burst observations, data analysis and interpretation in the context of multi-band and multi-messenger astrophysics and of observational strategies for the detection of the electromagnetic counterparts of gravitational wave sources. Member of the LIGO/Virgo Scientific Collaboration awarded of the 2016 Special Breakthrough Prize in Fundamental Physics for the first detection of a gravitational wave signal. Member of the ESA Science Study Team for the THESEUS space-based mission project for transient high-energy sky and early Universe surveys. Since 2017 she has joined the Cherenkov Telescope Array collaboration.

Andrea Rossi researcher at the INAF-OAS Bologna, night astronomer at the Large Binocular Telescope (LBT) for INAF. Member of the STARGATE collaboration for the follow-up of Gamma-ray bursts (GRB), and of GRAvitational Wave Inaf TeAm Collaboration (GRAWITA) for the follow-up of the electromagnetic counterparts of Gravitational waves sources (GW). His interests include: multiwavelength observations of host galaxies of GRBs and GWs, modelling of lightcurves of GRBs, supernovae, kilonovae, modelling of galaxy spectral energy distribution, surveys of galaxies.

Maria Giovanna Dainotti American Astronomical Society Chretienne Fellow at Stanford University and Assistant Professor at Jagiellonian University, affiliated scientist at Space Science institute in Colorado and Knight of the Italian Republic for scientific merit. Affiliated member of the Fermi Group for the study of high energy GRBs. Her interests include: multiwavelength observations of GRBs with particular attention to X-rays and gamma rays, study of GRB correlations, how to overcome the problem of selection bias, application of GRBs as cosmological tools, application of machine learning for redshift extraction of GRBs, modelling of GRBs in particular for the interpretation of the plateau emission within the magnetar scenario.

Preface to "Observations and Theory of Short GRBs at the Dawn of the Gravitational Wave Era"

The present issue contains a useful structured sample of articles that effectively summarize the status of the art on main aspects concerning gamma-ray bursts and binary neutron star mergers in light of GW170817. The current established knowledge and criticalities on numerical simulations of binary neutron star mergers and simulations of short GRB jets is reviewed in the framework of numerical relativity simulations of jet and magnetized outflow produced after merger. The issue then tackles the crucial question of the rate and detectability of GW170817-like events during the second generation gravitational wave network, as well as the main properties of the expected host galaxies to be targetted in the follow-up campaigns of the poorly localized gravitational wave sources. Although the multimessenger view of the short GRB170817 enabled us to gain deep insights on the physics of the jet emission mechanisms, open questions as the physical interpretation of the prompt emission of short and long GRBs and the early emission from the reverse shock are still to be answered and are here discussed.

Giulia Stratta, Andrea Rossi, Maria Giovanna Dainotti
Special Issue Editors

Review

Binary Neutron Star and Short Gamma-Ray Burst Simulations in Light of GW170817

Antonios Nathanail

Institut für Theoretische Physik, Goethe Universität Frankfurt, Max-von-Laue-Str.1, 60438 Frankfurt am Main, Germany; nathanail@th.physik.uni-frankfurt.de

Received: 16 August 2018; Accepted: 13 November 2018; Published: 19 November 2018

Abstract: In the dawn of the multi-messenger era of gravitational wave astronomy, which was marked by the first ever coincident detection of gravitational waves and electromagnetic radiation, it is important to take a step back and consider our current established knowledge. Numerical simulations of binary neutron star mergers and simulations of short GRB jets must combine efforts to understand such complicated and phenomenologically rich explosions. We review the status of numerical relativity simulations with respect to any jet or magnetized outflow produced after merger. We compare what is known from such simulations with what is used and obtained from short GRB jet simulations propagating through the BNS ejecta. We then review the established facts on this topic, as well as discuss things that need to be revised and further clarified.

Keywords: binary neutron stars; short gamma-ray bursts; GW170817

1. Introduction

The detection of GW170817 marked the dawn of the multi-messenger gravitational-wave era [1,2]. The subsequent observation of a short gamma-ray burst (GRB) almost ∼1.7 s after merger [3,4] showed that a least a subset of short GRBs is produced by binary neutron star (BNS) mergers. Hours after merger, a precise localization was established through optical observations of GW170817 [5,6], identifying the host galaxy as NGC 4993, which is at a distance of 40 megaparsecs (Mpc). Further detection in UV/optical/Infrared established the perennial connection between BNS mergers and a kilonova (macronova) [6–21].

A coincident detection of a GW and a short GRB from a BNS merger was long ago conjectured to be from short-duration GRBs come from BNS mergers [22–24]. These unprecedented observations open new windows and insights for the detailed study of such objects and events. These observations also opened the possibility of constraining the maximum mass of neutron stars and the equation of state (EOS) [25–39].

It was proposed some time ago that a BNS merger would give rise to emission powered by the radioactive decay of r-process nuclei [40,41]. Several groups concluded that this was the case for the optical/NIR emission that followed GW170817 [11,12,14,18,42–48]. This observation triggered further modelling for the actual components that give rise to this emission and how these components were produced.

The prompt gamma-ray emission was reported in [3,4]. It was the faintest (short or long) GRB ever detected [3]. The first X-ray afterglow observations came nine days after merger [19,49,50]. The first radio counterparts came later, sixteen days after merger [51,52]. All information that would come from the afterglow observations would be invaluable to reveal the nature of the outflow and its structure. A relativistic outflow from a BNS merger was indeed observed [52,53]. Was that the most peculiar short GRB ever detected [14,54]? The continuous rising of the afterglow the first 100 days suggested that a simple top-hat[1] jet model seen off-axis is not adequate for explanation [52,55–59]. However, at a 100 days after merger, the data could not exclude other jet structure or cocoon models. Energy injection was evident at that time [60,61]. Then, a turnover in the light curve appeared after nearly 200 days [62–64]. This emission is well understood and comes from the interaction of the outflow as it smashes into the inter-stellar medium producing a shock which accelerates electrons that radiate synchrotron radiation and can give a great insight in the whole structure of the initial outflow.

The expected number of BNS mergers from LIGO/Virgo in the years to come is 1–4 detections per year [65]. To digest all these new insightful observations, and those yet to come, we have to combine all the available data. What has been achieved from BNS numerical relativity simulations has to be part of any adequate modelling of short GRB outflows. These outflows are: the ejected matter and the production of neutrino-driven winds, the enormous magnetic field evolved in the merger process and its amplification during merger, and the actual possibility of launching a relativistic outflow after merger, which are the starting points given by numerical relativity. Is it a stable magnetar or the collapse to a black hole (BH) torus system that powers an outflow? In what follows, we try to present results from numerical relativity BNS simulations relevant for short GRBs. Afterwards, we turn our attention to efforts in short GRB jet simulations propagating through the BNS ejecta.

This is a rather focused review on what we know from numerical relativity concerning short GRBs and how this knowledge is applied to short GRB simulations. It will not at all follow the path of excellent reviews that exist on the subject of BNS mergers. For the interested reader, we cite several detailed reviews of subjects relevant to the detection of a BNS merger, a short GRB and a kilonova. Detailed reviews of all the aspects of numerical relativity and its applications to BNS mergers are given in [66,67], a focused review on BH–neutron star binaries is given in [68], review on the connection between BNS mergers and short GRBs in numerical relativity results are found in Refs. [69,70], observational aspects of short GRBs and connection to BNS mergers are reviewed in [71,72], the BNS merger and electromagnetic counterparts from kilonova are reviewed in [73–76], a review of rotating stars in relativity with applications on the post merger phase is given in [77], and reviews of short GRBs and entire aspects of GRBs are given in [78–82], respectively. In Section 2, we review the relevant knowledge from BNS simulations. We mainly follow results from magnetohydrodynamic (MHD) simulations in BNS studies. At the end of Section 2, we show the different paths that a BNS may follow after merger with respect to the achieved magnetic energy growth during merger. This translates to the total mass of the binary. In Section 3, we follow the studies that focus on the interaction of a BNS relativistic outflow passing through the matter that has been ejected during merger. In Section 4, we present the conclusions.

2. BNS Numerical Simulations

Sixteen systems of double neutron stars have been observed in our Galaxy. The observational data for the total mass of double neutron stars from our Galaxy show a narrow distribution in the range 2.58–2.88 $M_\odot$ [83]. A double neutron star system will inspiral and emit gravitational waves that result in orbital decay, shrinking their separation. When they come close enough, tidal forces result in deformation of the shape of the two neutron stars. Only numerical relativity can adequately describe the inspiral process beyond this point.

When the two neutron stars come into contact with one another, a merger product is formed. If this is massive enough and cannot support itself against gravitational collapse, a BH is formed in the first millisecond after merger, surrounded by a negligible disk. If the configuration is less massive, it can live longer. The merger product is differentially rotating and thus it can support more mass than the limit for a uniformly rotating star. At this stage, the merger product is called a hypermassive neutron

---

[1] A top-hat jet is one with constant Lorentz factor and emissivity within the jet that goes sharply to zero outside of jet opening angle. It is the simplest model to explain GRBs have been widely used to explain GRB properties.

star (HMNS) [84]. Gravitational-wave emission and magnetic field instabilities can remove angular momentum and make the HMNS unstable. The loss of thermal pressure due to neutrino cooling could also trigger the collapse of the HMNS [85,86], see also [87] for a slightly different conclusion. Moreover, if the total mass of the object is smaller than the mass that can be supported when allowing for maximal uniform rotation—the supramassive limit—it can also lose differential rotation and not collapse. This would result in a uniformly rotating supramassive neutron star (SMNS) surrounded by a disk. The SMNS will continue to loose angular momentum through magnetic spin down and also accrete mass from the surrounding disk. Its lifetime varies from a second to millions of seconds, in the latter case it can be considered as a stable configuration.

In the last years, a robust picture has been drawn regarding the ejected matter during and after merger from numerical simulations. These include matter ejected dynamically during merger and secularly after merger, such as neutrino driven winds and magnetic winds [88–108]. Other important properties of the merger product such as the spin and the rotation profile have been studied [109–111]. We continue focusing on the properties of the magnetic field, its amplification during merger and all the variety of observational outcomes that depend on the collapse time of the merger product and are dictated by the magnetic field.

Magnetic Field Amplification. The importance of the Kelvin-Helmholtz (KH) instability in BNS mergers was pointed out by Rasio and Shapiro [112]. As the stellar surfaces come into contact, a vortex sheet (shear layer) is developed which is KH unstable. The first simulation reporting on the KH instability for BNS is reported in [113]. It is reported in [114] that the KH instability could amplify the magnetic field beyond the magnetar level. They reported a lower value of 2×10^{15} G. However, they mentioned that numerical difficulties do not allow to reach the realistic values of amplification, which could be far above this limit. To address the full problem in numerical relativity is not so easy because high-resolution simulations are necessary, since the KH instability growth rate is proportional to the wave number of the mode, the shortest wavelengths grow the most rapidly. Studies of BNS mergers tried to clarify the picture and indeed showed some amplification, yet the saturation level was not pinpointed [115–121].

Another approach is local simulations that imitate the conditions of shear layers and study in detail the different phases of such a procedure. The growth phase where the KH vortex is formed, the amplification phase where the magnetic field is wound up by the evolving KH vortex, and the last phase where the magnetic field has locally reached equipartition that results in the KH vortex to lose its energy. In Figure 1, such a configuration is depicted after the end of the amplification phase. The blueish regions in the lower panel of Figure 1 indicate strongly magnetized regions that occur after amplification. Local simulations do not have such stringent resolution limitations as global simulations [122,123].

A high resolution study by Kiuchi [124] showed that, for an initial maximum magnetic field of 10^{13} G, the maximum magnetic field during merger and in the first 4–5 ms can reach 10^{17} G. They showed that the saturation magnetic energy is above $\gtrsim 4 \times 10^{50}$ erg, which is $\gtrsim 0.1\%$ of the bulk kinetic energy. Going to even higher resolution and running for a longer time, the upper bound for the amplified magnetic energy has not been reached yet. Higher values of the amplified magnetic energy live in denser regions [125]. This may indicate that the higher values of the magnetic field are either trapped in the dense core, or that they need a diffusion timescale to diffuse out from the core and reorder [126]. These results have built stable foundations that magnetic field amplification is an integral part of the BNS merger and happen in the first millisecond after merger, as seen in Figure 2. Another important point to make here is that this is true only if the binary does not experience a prompt collapse, in which case there is no time to amplify the magnetic field and the EM output of the remnant follows a different path. We focus on this in more detail later.

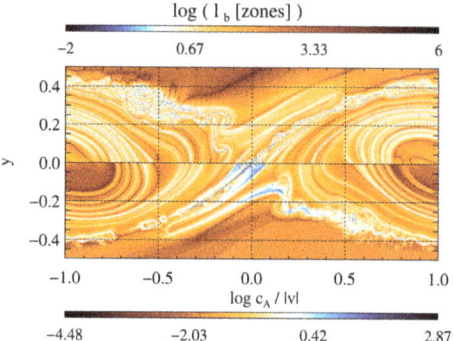

Figure 1. Snapshots of a certain model from [122]. It is taken shortly after the termination of the kinetic amplification phase. The top panel shows the characteristic length scale of the magnetic field, $|B|/|\nabla \times B|$ in units of the zone size. Regions where magnetic structures are larger than one computational zone are depicted in orange-red colors and blue colors where they are smaller than one zone size. The bottom panel shows the ratio between the Alfvén velocity and the modulus of the fluid velocity. Strongly magnetized regions are depicted in blue, whereas weakly magnetized regions are depicted in orange-red. (Reproduced with permission from [122], © ESO, 2010).

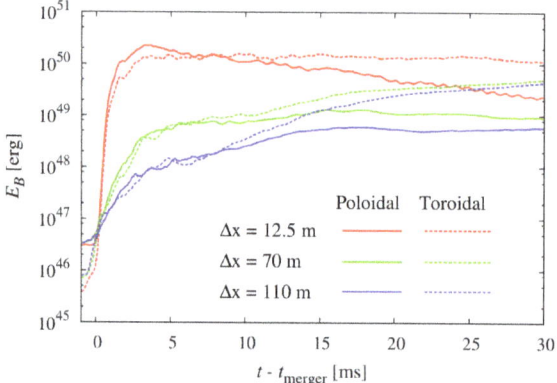

Figure 2. The evolution of the magnetic-field energy as a function of time from [125]. The growth of the magnetic field is evident in the first five milliseconds. However, the strong dependence on resolution indicates that the upper limit of amplification is unknown. Solid and dashed curves indicate the poloidal and toroidal magnetic field components, respectively. (Reprinted with permission from [125]. © (2018) by the American Physical Society.)

Another cause for magnetic field amplification is magnetic winding due to differential rotation, which continues to function even after the KH instability may saturate. Furthermore, there are indications from studies of core-collapse supernovae that the magneto-rotational instability (MRI) can also be important [127]. From such simulations, it has been shown that the MRI can amplify the magnetic field by a factor of 100. The importance of parasitic instabilities that may quench such mechanisms have also to be taken into account [128].

Observational signatures during magnetic field amplification. Are there direct observational signatures of the field amplification? The magnetic energy increases to extreme values. It has been proposed that if only a fraction of this energy dissipates through reconnection it yields an EM counterpart at the time of merger. This could be observable up to a distance of 200 Mpc [129]. This radiation can only escape if produced in an optically thin surface layer. However, the higher values for amplification were reported in the dense core of the merger remnant [125]. The evolution of this turbulent magnetic field is not yet fully understood, and it may take a much longer time than the merger timescale to diffuse out of the dense core [126]. If the merger remnant lives for at least a second, then the Hall effect becomes important, and would govern the structure of the magnetic field at late times [126].

BH torus from BNS in MHD. Strong magnetic fields are present during and after the merger of a BNS. The next meaningful ingredient is the outcome and lifetime of the merger remnant. Due to numerical limitations, existing studies cover the collapse of the merger remnant to a BH only if it happens prior to ∼100 ms after merger. It was long ago proposed that BNS mergers could launch a short GRB. This connection had been made clear by recent observations [52,53]. However, it is still something yet to be achieved by global simulations. The first attempts in a magnetohydrodynamic (MHD) setting and in full GR did not show any signs of jet production following merger and the subsequent collapse of the merger remnant [115,116]. In subsequent studies, a magnetic jet structure was reported as a low density funnel with ordered poloidal magnetic field above the BH [130]. This is indeed the first step to imagine the production of a relativistic magnetized jet. Another important aspect is that an ordered poloidal magnetic field is needed to account for energy extraction from the BH in a Blandford-Znajek framework [131]. However, even if the magnetic field is not poloidal there could be other outcomes for an outflow. Another simulation by a different group did not find such a structure, instead the conclusion drawn from their simulations indicated an expanding toroidal field [121], which is also capable of producing a jet configuration with a different underlying mechanism [132].

It was further shown and confirmed in a resistive MHD framework that, at least for merger remnants that collapse in the first ∼10 ms, the BH-torus system produces a low density funnel above the BH [119]. The excess of the internal energy in this low density region above the newly formed BH could lead to the production of a jet (Figure 3). However, how low is low? To launch an outflow, it is necessary that at least the magnetic pressure in the jet interior can accelerate the fluid in the polar region. Previous studies [119,130] used an ideal fluid equation of state (EOS), whereas in [121], a piece-wise polytrope was used, and it was pointed out in [69,133] that the jet structure indeed depends on the EOS. Other studies have also reported the production of a magnetic structures when using a different EOS [134,135], also including a neutrino treatment [120]. In Figure 4, such a configuration with a BH-torus system is depicted. In this specific model, the merger product collapsed to a BH at t_{BH} ∼8.7 ms. The snapshot is taken at t ∼35.1 ms after merger. In the low density funnel above, the BH the magnetic structure is clearly seen.

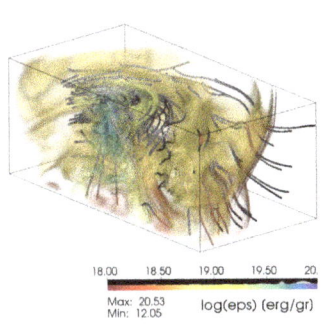

Figure 3. The structure of the torus for a model from [119]. 3D snapshot of the specific internal energy and the magnetic field lines at t = 18.3 ms. Two main points are illustrated in the figure: (i) the structure of the toroidal magnetic field inside the torus; and (ii) the excess of internal energy close to the polar axis where the low density funnel is produced. (Reprinted with permission from [119]. © (2016) by the American Physical Society).

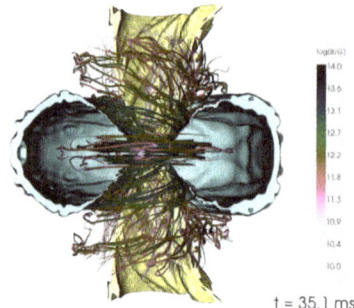

t = 35.1 ms

Figure 4. The magnetic field structure for a model from [134] depicted at 35.1 ms after merger. Two isosurfaces of density are shown in yellow (10^8 g/cm^3) and cyan 10^{10} g/cm^3). The field lines are colored by magnetic field strength. The toroidal field inside the torus is easily seen, together with a poloidal funnel above the BH. This model collapsed to a BH at $t_{BH} \sim 8.7$ ms after merger. Due to the limited resolution, the KH instability is not entirely accounted for in these simulations. (Reprinted with permission from [134]. © (2016) by the American Physical Society).

Recently, a production of an incipient jet (as termed by the authors) was reported, which attained a Poynting luminosity of $\sim 10^{51}$ erg/s and a maximum Lorentz factor of $\Gamma = 1.25$ [133]. Towards the end of the simulation, they reported a magnetically dominated funnel above the BH, which can be seen in the lower panel of Figure 5. The snapshot is taken at $t \sim 67.7$ ms, whereas the merger product collapsed to a BH at $t_{BH} \sim 18$ ms after merger. It is clear that at late times the low density funnel above the BH is decreasing even more rapidly in density. This allows a magnetically dominated region to evolve. Using the magnetization of the outflow, they estimated the half opening angle of the jet funnel to be ~ 20–$30°$ [133].

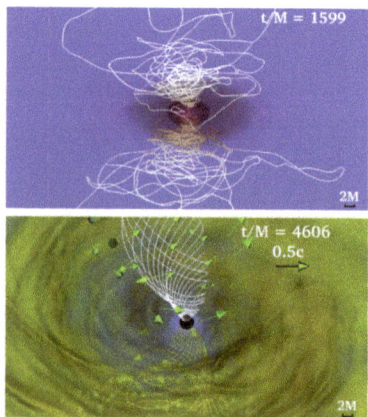

Figure 5. Snapshots of the rest-mass density of a model from [133]. Magnetic field lines are depicted as white lines and arrows indicate plasma velocities. In this model, the merger remnant collapses to a BH at $t_{BH} \sim 1215$ M = 18 ms after merger. The upper panel is at a slightly later time after collapse, whereas the lower panel is at $t \sim 67.7$ ms. We point out that, while the density contours are selected far from the magnetic jet structure, the funnel is filled with low density matter which supports the collimation of the magnetic structure. The length scale of the plots is $M = 4.43$km. (Reprinted from [133]. © AAS. Reproduced with permission).

EM luminosity. It is natural to ask why there is so much discussion about magnetic fields and their role in the production of jets. Other mechanisms have been proposed, such as neutrino annihilation [22,136]. However, recent studies in which neutrinos are also treated to study a BNS merger and the evolution of accretion to a BH, it was found that due to a highly baryon-loaded environment such a mechanism alone does not suffice [47,137]. On the other hand, the electromagnetic energy extraction from a BH (the BZ mechanism) has been widely studied (numerically [138,139], semi-analytically [140] and analytically [141]) and widely understood and accepted. It needs only two ingredients, a rotating BH and an ordered poloidal magnetic field to extract this rotational energy.

$$L_{BZ} \sim \frac{1}{6\pi^2 c}\Psi_m^2 \Omega_{BH}^2 \sim B_p^2 R_{BH}^2 \left(\frac{\alpha}{M_{BH}}\right)^2$$
$$\sim 10^{51} \left(\frac{B_p}{2\times 10^{15}\text{G}}\right)^2 \left(\frac{M_{BH}}{2.8 M_\odot}\right)^2 \left(\frac{\alpha}{0.8 M_{BH}}\right)^2 \text{erg s}^{-1}, \quad (1)$$

where Ψ_m is the magnetic flux accumulated on the BH horizon, Ω_{BH} is the angular velocity of the BH, B_p is the poloidal magnetic field on the BH horizon, α is the spin parameter of the BH and M_{BH} is the mass of the BH [142].

In a BNS merger, one has both: when the merger remnant collapses a BH is formed, and in all reported cases it attains a spin of ~0.8. The magnetic field is known to be an essential ingredient of a neutron star and as we have already discussed it is further amplified during merger. In a baryon-polluted environment such as the one that exists around the remnant after merger, there are also other things to worry about. The ram pressure from the material from the polar regions, or even fall-back material in this region, may not allow this outflow to form and evolve. This is perhaps the reason, together with a low magnetic field, that in some studies with limited amount of evolution time no outflow was formed [134]. If this is the case, then it is expected that some hundreds of milliseconds later the overall pressure of the funnel could decrease significantly, allowing for a magnetically dominated outflow to emerge.

Duration of a BH torus. Following the above discussion, it is natural to ask how long this configuration will last. This is indicated by the mass of the surrounding disk plus the mass accretion rate. We briefly discuss the duration connected with the mass of the torus. It is usually assumed that the duration of the short GRB (<2 s) is due to the accretion timescale of the surrounding torus. Studies have shown that the mass of the torus can be as large as $M_T \sim 0.001$–$0.2~M_\odot$ [108,115,134,143–146]. Through numerical simulations a simple phenomenological expression can be derived that reproduces the mass from the surrounding torus [91,144]. A general result is that unequal mass binaries have a more massive torus around the BH that is formed. On the other hand, equal mass binaries acquire less massive torii. Of course, in the case of prompt collapse, the surrounding torus is negligible, but this is something we discuss after commenting on the accretion timescale. Furthermore, in the case of a late collapse the surrounding disk is expected to be negligible [147,148].

The duration of any event coming from the BH torus depends on the lifetime of the torus, and this torus will persist on an accretion timescale. A rough estimate for the viscous accretion timescale can be given as:

$$t_{accr} \simeq 1 \left(\frac{R_T}{50~\text{km}}\right)^2 \left(\frac{H_T}{25~\text{km}}\right)^{-1} \left(\frac{\alpha}{0.01}\right)^{-1} \left(\frac{c_s}{0.1c}\right)^{-1} \text{s}, \quad (2)$$

where R_T and H_T are the radius and the typical vertical scale height of the torus, c_s is the speed of sound and α is the α-parameter that describes the efficiency of angular momentum transport due to turbulence in the torus [149]. As such, if the BNS merger produces a BH torus system, the accretion timescale sets the duration of the outflow, if any outflow is produced. However, we note here that it is also important to discuss the duration of a gamma-ray pulse produced by a relativistic outflow in a different fashion. The photosphere is defined as the radius that the outflow first becomes transparent and the first photons are emitted. If an outflow has attained a Lorentz factor Γ, then photons emitted at any point on the jet are beamed within a $1/\Gamma$ cone, as seen in the lab frame. Thus, assuming that the outflow has a conical shape with opening angle θ_j,[2] initially when $\Gamma > 1/\theta_j$, an observer can see only radiation from a small fraction of the jet. The duration of the pulse can be interpreted as photons coming from this cone that the observer is able to see, the $1/\Gamma$ cone. For a mildly relativistic outflow with $\Gamma > 1/\theta_f$, the relevant timescale of the pulse is

$$dt \sim 1-2 \left(\frac{r_{em}}{10^{12} \text{cm}}\right) \left(\frac{\Gamma}{6-10}\right)^{-2} \text{s}, \quad (3)$$

where r_{em} is the emission radius [80]. The key point here is that, even if the accretion timescale is shorter and a relativistic outflow is produced, the duration can also be explained by other robust physical arguments. For an ultra-relativistic outflow, the duration of the pulse is very small and as such the accretion timescale can enter as a justification of the duration of the event.

The discussion thus far is mainly for a merger remnant that collapses to a BH after 10 ms or more. The effect of the collapse of the merger remnant when it occurs in the first milliseconds is different. The general thinking in the community leads to no expectations for an EM counterpart, if the BNS merger undergoes a prompt collapse to a BH. This is based on results of simulations that showed some robust features of this evolution track in the case of an equal mass binary. These features show that a limited amount of mass is dynamically ejected, and thus no expectation whatsoever of a kilonova. Another feature is the limited amount of time between merger and collapse, which prohibits significant magnetic field amplification, and as a result the magnetic energy will not reach such large values. However, a detailed high-resolution study of a prompt collapse has not yet been performed.

Lastly, the limited amount of mass left around the BH cannot sustain any magnetic structure for longer than a few milliseconds. This means that whatever is formed after merger will be lost on this timescale. However, the magnetic field that remains outside the BH will dissipate away on this timescale. Most of the matter will be lost behind the BH horizon, but the magnetic field lines will snap violently. This will produce a magnetic shock that dissipates a significant fraction of the magnetic energy by accelerating electrons, producing a massive burst, similar to a blitzar [150]. This can produce an EM counterpart on such a timescale. Prompt collapse events produce less massive accretion disks than those arising from delayed collapse. Studies have shown that the result of a prompt collapse is a spinning BH and an accretion disk with a negligible mass of $M_T \sim 0.0001$–$0.001\ M_\odot$ [115,143,144,151–153]. A negligible mass for the surrounding torus in the delayed collapse scenario can of course also be due to the underlying EOS [91,144].

Prompt Collapse. The prompt collapse also has an impact on the magnetic field evolution. Since the HMNS lifetime is limited, the magnetic field amplification is also limited [30]. However, a precise value for this upper limit is not known. The mass threshold at which the HMNS promptly collapses to a BH strongly depends on the EOS [144,152,154–156]. It is clear that a soft EOS, meaning that matter can be compressed in a more effective way, is more compact and the threshold mass to collapse to a BH is smaller. Conversely, a stiff EOS does not allow for such compression and a star is less compact, and therefore allowed to have a larger threshold mass [144]. A BNS with a total mass of $M_{tot} \sim 2.8\ M_\odot$ can in principle promptly collapse to a BH, whereas for a slightly less massive system it can lead to a delayed collapse some milliseconds after merger [152]. Reducing even further the total mass to be $\lesssim 2.7\ M_\odot$, a stable configuration can be achieved. Interestingly, from the known double neutron star systems observed in our Galaxy, the total mass is around $\sim 2.7\ M_\odot$ [157]. This means that we could expect all outcomes: i.e., prompt collapse, delayed collapse or a stable configuration.

It was reported that following a prompt collapse to a BH no kind of jet can be formed [158]. The system does not have the time to develop a jet structure. However, it possesses a magnetic field for which we do not know precisely the level of amplification. When the negligible torus

---

[2] An outflow that has a finite angular extent that ends at the boundary of a cone has an opening angle defined by the axis of the cone and the cone itself.

is eventually accreted, all this magnetic energy will be dissipated away. As discussed previously, prompt collapse also leads to a very small torus. The torus lifetime can be as small as $t_T \sim 5 \left(\frac{M_T}{0.001\,M_\odot}\right)\left(\frac{\dot{M}}{0.2\,M_\odot\,s^{-1}}\right)^{-1}$ ms [158]. We may estimate the energy stored in the nearby magnetosphere to be

$$E_{EM} \simeq 10^{40}\, b_{12}^2\, r_{10}^3 \text{ erg}, \qquad (4)$$

assuming no amplification has taken place. It has a millisecond duration and an energy close to the requirement for a fast radio burst (FRB [159,160]). Overall, this could be similar to the model proposed for FRBs where a supramassive neutron star collapses to a BH [150,161]. Thus, a prompt collapse is lacking many interesting features arising from the delayed collapse, but could provide answers to other mysterious EM signals (see also [162]). We must add that in the event that the magnetic field energy is amplified to above 10^{47} erg in the first millisecond after merger and the remnant subsequently collapses to a BH, the interaction of the emergent magnetic pulse with the ejected matter could give rise to a different variety of low luminosity short GRBs.

SMNS spin down. A stable neutron star configuration may also be the end point of a BNS merger. If the total mass of the binary is below a certain limit, then even significant accretion of the surrounding matter cannot trigger its collapse. This may have distinct observational features and could explain X-ray plateaus in the afterglow of short GRBs [163]. It has been suggested that a long-lived magnetar as a BNS merger product can power such emission by its spin down dipolar radiation [164–168]. Such simulations showed that a stable neutron star with a surrounding disk can be a BNS merger product and the luminosity from such a configuration is significant [169,170]. However, the first gamma-rays from the short GRB could not be explained. To overcome this drawback, different scenarios have been proposed. The production of the gamma-rays is attributed to the collapse of this long-lived object to a BH, which happens after the production of the X-ray radiation. The observational features of such a model, together with the prompt gamma-rays of a short GRB, come from diffusion arguments [171,172].

In most studies, the long-lived SMNS is losing angular momentum due to magnetic spin down and the production of dipolar radiation where energy is lost at a rate

$$\dot{E}_{mag} = \frac{\mu^2 \Omega^4}{c^3}(1 + \sin^2 \chi), \qquad (5)$$

where $\mu = B r_{NS}^3$ is the magnetic dipole moment, B is the dipole magnetic field, r_{NS} is the neutron star radius, Ω is the angular velocity and χ is the inclination angle between the magnetic and the rotation axis [173,174]. However, if a stable object is produced, it lives entirely in the environment of a surrounding torus starting exactly at the surface of the star [169]. This means that it is impossible for this neutron star to acquire a dipolar magnetic field, since the magnetic field loops cannot close through the torus, but have instead opened up either during merger or due to differential rotation [98]. Additionally, if any closed field lines remain, they are influenced by neutrino heating [175–177]. However, this effect will be lost in 1–2 s. The last, but most significant, argument is that field lines which thread the disk will open up, due to the differential rotation of the two footpoints of the magnetic field line, one anchored on the SMNS and the other threading the disk, similar to the BH case [178–180]. Even if most of the mass of the disk is accreted or expelled, the remaining negligible mass will not allow the field lines to close. Thus, the structure of the magnetosphere of the merger remnant can be approximated by a split-monopole configuration [27]. A neutron star with a split-monopole configuration spins down with a different dependence on rotation, similar to a BH spin down where all field lines are also open [181,182]. The spin down follows an exponential decrease

$$\dot{E}_{mag} = -\frac{2}{3\pi c} B^2 r_{NS}^4 \Omega^2 \exp^{t/\tau_B} \simeq 5 \times 10^{50} \left(B/10^{15}\,G\right)^2 \\ (r_{NS}/12\,\text{km})^4 (P/1\,\text{ms})^{-2} \exp^{t/\tau_B} \text{ erg s}^{-1}, \qquad (6)$$

where $\tau_B = 67 \left(B/10^{15}\,\mathrm{G}\right)^{-2} (r_{NS}/12\,\mathrm{km})^{-2}\,\mathrm{s}$. Essentially, the spin down of such a configuration progresses more rapidly than the dipolar case, since all the field lines are open and contribute to the spin down process.

SMNS and the surrounding disk. The next thing that we want to focus is the evolution of the disk that surrounds the SMNS and the outcome of the collapse of the SMNS after one second from merger. Due to transfer of angular momentum the disk expands over time and due to accretion onto the compact remnant, its mass decreases over time [108]. As in Equation (2) the viscous accretion timescale estimated for the torus:

$$t_{accr} \simeq 1\,\mathrm{s} \left(\frac{R_T}{50\,\mathrm{km}}\right)^2 \left(\frac{H_T}{25\,\mathrm{km}}\right)^{-1} \times \left(\frac{\alpha}{0.01}\right)^{-1} \left(\frac{c_s}{0.1c}\right)^{-1}, \quad (7)$$

where H_T is the typical vertical scale height of the torus and R_T is its radius. Therefore, the mass accretion rate onto the SMNS yields

$$\dot{M}_{SMNS} \simeq \frac{M_T}{t_{accr}} \sim 0.2\,M_\odot \mathrm{s}^{-1} \left(\frac{\alpha}{0.01}\right) \left(\frac{M_T}{0.2 M_\odot}\right) \times \left(\frac{R_T}{50\,\mathrm{km}}\right)^{-2} \left(\frac{H_T}{25\,\mathrm{km}}\right), \quad (8)$$

where M_T is the mass of the torus. The mass of the torus decreases in time as the torus expands, thus this accretion rate is not stationary. This effect is seen in Figure 6, where the density profile is shown in the equatorial plane at different time slices. As time goes by, the torus expands and the torus density decreases significantly. The radius of the torus may reach 140 km in 1 s. The total mass accreted can be estimated to be $\sim 0.12\,M_\odot$ in 1 s [108].

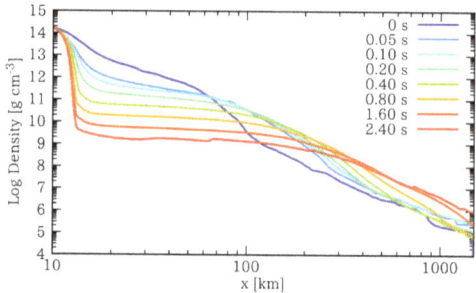

Figure 6. Density profiles on the equatorial plane at different time slices $t \sim 0$, 0.05, 0.1, 0.2, 0.4, 0.8, 1.6, and 2.4 s. The torus gradually expands with time and its density decreases. This is due to viscous angular momentum transport. Even one second after merger the SMNS resides in a low density torus, in contrast to its inherent nuclear densities. (Reprinted from [108]. © AAS. Reproduced with permission).

If the SMNS is close to its maximum mass limit, this significant mass accretion in one second may trigger its collapse. Furthermore, the expansion of the torus is also significant during this time.

The density of the torus in the vicinity of the SMNS could designate the outcome of the collapse to an induced magnetic explosion. The estimation for the density of the torus at 1s yields:

$$\rho_T \simeq \frac{M_T}{2H_T \pi R_T^2} \sim 9.2 \times 10^9 \, \text{g/cm}^3 \left(\frac{M_T}{0.08 M_\odot}\right) \times \left(\frac{R_T}{140 \, \text{km}}\right)^{-2} \left(\frac{H_T}{70 \, \text{km}}\right)^{-1}, \quad (9)$$

where quantities are for the expanded torus at 1 s after merger. The density in the poloidal plane is shown in Figure 7 at time $t \sim 1.6$ s after merger. The density drops around 5–6 orders of magnitude in the first 1300–1500 km. The possibility that no debris disk is formed at all has also been discussed [147,148].

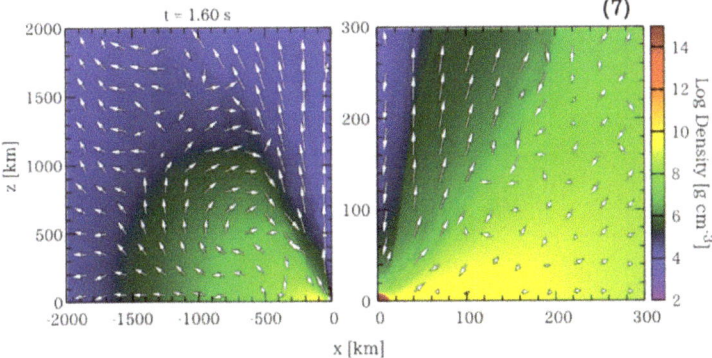

Figure 7. Snapshots of the density and poloidal velocity field for a model from [108]. The velocity vectors and their length correspond to the logarithm of the velocity in the poloidal plane. The left panel shows a region of 2000 km, whereas the right panel a narrower region of 300 km. This profile corresponds to $t \sim 1.6$ s after merger. (Reprinted from [108]. © AAS. Reproduced with permission).

Table 1. Outcome of the collapse of the merger remnant, the different columns indicate the different possible outcomes for the merger remnant. The different outcomes depend on the collapse time to a BH. Different rows, from top to bottom, are: the collapse time to a BH t_{BH}, if magnetic field amplification occurs or not, the amount of magnetic energy E_B, if there is ejected matter, the amount of mass surrounding the BH when it is formed, the lifetime of this disk around the BH, whether the EM outcome will be produced either by collapse or by the absence of collapse, and the estimated energy that is released during the collapse or the absence of collapse.

Possibilities for the Merger Remnant	Prompt Collapse	Delayed Collapse	"Further" Delayed Collapse	No Collapse
collapse to BH, t_{BH}	1–2 ms	7–500 ms	1–3 s	∞
B-amplification	not significant	yes	yes	yes
Magnetic energy, E_B	10^{40}–10^{44} erg	10^{51} erg	10^{51} erg	10^{51} erg
ejecta	not significant	yes	yes	yes
BH surrounding disk	negligible	0.05–0.2 $M_\odot$	0.01–0.05 $M_\odot$	no BH disk
disk lifetime	2–8 ms	0.2–1 s	0.1–0.2 s	0
EM outcome	magnetic energy dissipation	magnetic jet	magnetic explosion	magnetic wind (spin down)
Estimated energy	10^{40}–10^{44} erg	10^{51} erg	10^{51} erg	10^{50} erg

Jet or magnetic explosion. Previously, we discussed the production of a low density funnel that appears after the collapse of the merger remnant to a BH. All results from simulations so far describe such an evolution in the case that the collapse occurred in the first milliseconds after merger. Here, we describe the conditions and the outcome of the collapse to a BH, if this happens after 1 s from merger. The foremost point is the condition for the establishment of a magnetic jet. A stable magnetic jet configuration needs the torus pressure to balance the magnetic pressure from the jet itself. Due to magnetic field amplification discussed earlier, we assume that the mean magnetic field of the SMNS is $B \simeq 10^{16}$ G. This yields:

$$\frac{B_{SMNS}^2}{8\pi} \simeq 4 \times 10^{30} \, \text{dyn/cm}^2 \left(\frac{B_{SMNS}}{10^{16} \, \text{G}}\right)^2$$
$$\gg 9.2 \times 10^{29} \, \text{dyn/cm}^2 \left(\frac{\rho_T}{9.2 \times 10^9 \, \text{g/cm}^3}\right) \simeq \rho_T c^2. \tag{10}$$

At later times, the torus has expanded even more and the establishment of a magnetic jet becomes more problematic due to the imbalance between the magnetic pressure and the disk ram pressure. We may also use the accretion rate at 1s as reported in [108], which is $\sim 0.02 \, M_\odot \, \text{s}^{-1}$. This yields:

$$B_{SMNS}^2 / 8\pi \gg 2.6 \times 10^{28} \, \text{dyn/cm}^2 \sim \dot{M}c/4\pi r_{BH}^2.$$

Figure 8 summarizes the above discussion. The main point is that if the collapse is triggered around or after ~ 1 s after merger, the magnetic energy of the SMNS is released and induces a powerful explosion of $E_{exp} \sim 10^{51}$ erg, contrary to the expectations of a magnetic jet [183].

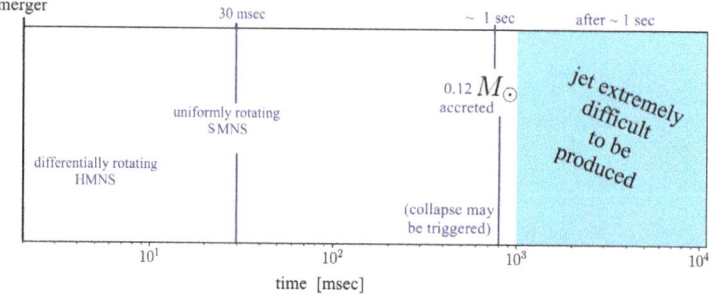

Figure 8. The lifetime of the merger remnant of mass close to the maximum for uniform rotation. The remnant does not collapse when differential rotation is lost, but collapse may be triggered when almost 0.1 $M_\odot$ has been accreted. If collapse is triggered after one second, then the production of a jet may not be favoured. In this case, an explosion is triggered, releasing the enormous amounts of magnetic energy stored in the magnetosphere of the SMNS as discussed in [183]. (Reprinted from [183]. © AAS. Reproduced with permission).

We may summarize the understanding of the outcome of the collapse of the merger remnant, which strongly depends on the time that the collapse is triggered. Of course, the triggering of the collapse depends on the EOS and the total mass of the binary, however we do not go to that great a depth here and instead characterize only the outcome with respect to the collapse time. The possible outcomes are summarized in Table 1. The four columns represent four different types for the outcome of a BNS merger. The different rows show characteristics that are essential to the observable outcome of a BNS merger.

The prompt collapse that is characterized by the collapse of the merger product in the first 1–2 ms (first column of Table 1) does not have an effective magnetic field amplification phase and also no significant ejecta, but due to the negligible disk that surrounds the newly formed BH the lifetime of this disk is on the order of few milliseconds. As a result, all its magnetic energy will dissipate on that timescale. The energetics of such an explosion (depicted in Equation (4)) and its timescale point to an event similar to FRBs. In all cases that the remnant lives longer than the first few milliseconds, it is certain that the magnetic field is amplified to high values. The case where the merger product (a HMNS at this stage) collapses in a few milliseconds to tens of milliseconds, is the most discussed case. This is expected to produce a canonical magnetic jet that interacts with the merger ejecta. If the collapse is delayed for a second (or more), then the low density of the torus may be insufficient to act as a boundary for a magnetic jet and a magnetic explosion is triggered.

At the end of this section, we list some interesting and critical points known from numerical simulations of BNS mergers and provide some comparison with points known from short GRBs.

Critical points:

- If the merger product does not collapse in the first millisecond, then the magnetic energy is amplified to values higher than 10^{50} erg [125].
- The saturation level of magnetic field amplification is not yet known [125].
- The amplified magnetic field is turbulent and requires time (more than a second) to rearrange in a coherent large-scale structure [126].
- After the collapse to a BH in 10 ms, a magnetic jet structure is produced [130,133,134].
- An ordered poloidal magnetic field above 10^{15}G is needed for a BZ luminosity of $\sim 10^{51}$ erg/s [131].
- The production of an ultra relativistic outflow has never been reported in BNS simulations [119–121,130,133,134].
- The magnetic jet funnel reported in BNS simulations has an opening angle of $\gtrsim 20$–$30°$, and a maximum Lorentz factor reported as $\Gamma = 1.25$ [133].
- If the collapse of the SMNS to a BH occurs late enough, the mass of the surrounding disk is negligible [147,148].

All these critical points should be taken into account for the understanding of any magnetized outflow (relativistic or non-relativistic) that emerges from the merger remnant or the collapse of the merger remnant to a BH. To help comparisons with observations, we should also mention here that there are short GRBs observed with a lower limit on the opening angle $\gtrsim 15°$ and some observed short GRBs that have jets with opening angles of 7–8°, [72]. However, the opening angle given from numerical relativity simulations at the base of the jet may (most probably) change through the interaction with the BSN ejecta. This is discussed in the next section.

3. Short GRB Jet Simulations

It is understood that if the merger does not follow a quick prompt collapse then significant mass is ejected following the BNS merger. Mass can be ejected dynamically, by winds driven from the newly formed HMNS and from the debris disk that forms around it [88–108]. As a result, any outflow that emerges from the merger remnant or the collapse of the merger remnant has to pass through this dynamical ejecta.

To continue further in the discussion of the interaction between the BNS ejecta and a (perhaps mildly) relativistic outflow that emerges after merger, we need to define characteristic names widely used in the literature. We follow the terminology as is clearly given by Nakar and Piran [184]:

It is important to define the angle with which the observer is looking at the emission produced from the outflow with respect to the motion of the outflow itself. Assuming that an emitting region moves relativistically with a Lorentz factor Γ, then the emission is termed:

On-axis emission: If the angle θ between the line-of-sight and the velocity of the emitting material satisfies $\theta \lesssim 1/\Gamma$. This emission is Lorentz boosted for relativistically moving material.

Off-axis emission: If the angle θ satisfies $\theta \gtrsim 1/\Gamma$. In this case, relativistically moving emitting material appears fainter than being on-axis. It is clear that emission, which originally is observed off-axis, will become on-axis when the emitting material decelerates significantly and expands sideways. Originally, on-axis emission always remains on-axis. We should also point out that the observer angle is usually defined as the angle between the jet axis (the symmetry axis) and the line-of-sight. For BNS mergers it is generally supposed that the jet axis coincides with angular momentum axis of the BNS system. Next, we define characteristic names concerning the intrinsic properties and structure of the emitting material.

Structured relativistic jet: As the name indicates, this is a relativistic jet along the symmetry axis that acquires a certain structure. This structure can be angular and/or radial. A simple example can be a "top-hat" jet, a blast wave where the energy and radial velocity are uniform inside a cone (Blandford-McKee [185]). Another example, usually inferred for short GRBs, is a successful jet with a cocoon, where the cocoon term is defined below. In general, a jet can be composed of a fast core at small polar angles surrounded by a slower, underluminous sheath. The presence of a spine-sheath structure can be independent from that of a cocoon.

Cocoon: If a jet propagates within a dense medium, then the jet transfers energy and shocks this material. There is also a reverse shock that goes down to the jet itself. The resulting configuration is called a cocoon. In the case of BNS mergers, the dense medium is the ejected material (dynamical and secular ejecta). Thus, if a jet is produced after merger, then a cocoon is also produced. There remains a differentiating factor of whether the jet was successful.

Choked jet with cocoon: The jet that produced a cocoon from the interaction with a dense medium did not have enough energy to break out of the medium and it is choked. The jet transfers all of its energy to the medium and the shocked material may acquire a certain angular structure. The reverse shock may also produce a radial structure inside the cocoon in the region of the choked jet. In the case of BNS mergers, a choked jet would mean that no usual short GRB was produced. However, a mildly relativistic outflow may be produced.

Successful jet with cocoon: The jet that produced a cocoon from the interaction with a dense medium had enough energy to break out of the medium. An ultra-relativistic outflow passed through the medium and eventually decelerates through the interaction with the inter-stellar medium (ISM). The jet transferred some of its energy to the medium and a cocoon was produced. In the case of a BNS merger, a successful jet would mean that a usual short GRB was produced, pointing along the jet (BNS) axis. However, a mildly relativistic outflow may also be produced. In this case, two components can be identified, an ultra-relativistic core which is surrounded by a mildly-relativistic cocoon.

Successful explosion (not jet): Assuming the possibility discussed in the previous section that a jet is never formed, we could rephrase the last case to a successful explosion with a cocoon. This means that no jet was formed but rather an instantaneous explosion occurred which followed the delayed (over a second) collapse of the remnant [183]. In such a scenario, the core is not ultra-relativistic, but just slightly faster than the surrounding cocoon itself.

It is important to note that there exist previous studies that have discussed the formation of cocoons in a slightly different context, namely, for long GRBs where the jet has to propagate through the stellar envelopes and not the BNS ejecta. The main differences should be in the density profile and how it falls off. Cocoons in long GRBs have been discussed in [186]. The mixing of the cocoon components has been discussed in [187–191].

In what follows, we review studies that have developed a robust picture regarding the outcome of a BNS merger with respect to the prompt emission which is a short GRB and the afterglow emission which can provide physical insight into the outflow that produced it. The common understanding for the prompt emission is that it is powered by some internal dissipation mechanism within the jet. The common interpretation of the late afterglow is that the interaction of the produced outflow with the ISM, during which the outflow sweeps up matter from the ISM, results in the eventual deceleration of the outflow.

Jet through the BNS ejecta. In this respect, Nagakura et al. [192] took into account the density profile of a BNS numerical relativity simulation [96] to study the propagation of a hydrodynamical jet through such ejecta and develop a picture of whether the jet could break out from them or not. Such studies built a consensus that even if the outflow emerging from the BNS has a wide opening angle, it will be subsequently collimated as it tries to pass through the ejecta[192–194]. These works described a density distribution that the jet should pass through, and that this density distribution of matter has been ejected primarily during merger. Any outflow produced in the base of the merger configuration has to pass through these ejecta and may change its shape through collimation or loose some energy by the interaction with the ejecta. This way, some energy deposits to the ejecta producing a cocoon structure.

In the work of [192], the jet opening angle was placed to be 15–45°, with an injected luminosity of $L \sim 10^{50}$ erg/s. As they pointed out, their results were similar to equivalent simulations in the context of the collapsar model [191,195]. The opening angle at the base of the jet is determined through the interaction of the jet and the surrounding disk. An important consequence of this study is the finding that irrespective of the initial opening angle, all jets succeed in breaking out and form what we would call a structured jet with a cocoon. Only for the model with an initial opening angle of 45° is this not the case, and a choked jet with cocoon is formed instead. Due to the large cross section of the jet, it cannot move sideways into the cocoon and expands quasi-spherically.

In Figure 9, a model from [192] is shown. The ejected mass is $10^{-3}\,M_\odot$ and the initial jet is injected with an opening angle of 15°. The density profile of the produced structure is shown for two snapshots, one at the time that the jet breaks out from the ejecta and the other at the end of the simulation. The average opening angle of the jet after break out, which has changed due to the interaction with the surrounding ejecta is $\theta_{jet} \sim 12.6°$. Interestingly, except the break out of the jet, a cocoon is formed and is clearly shown in the above mentioned Figure 9. However, there does not exist in this study a detailed description of this component. The density profile for the ejecta used in this study has a steep profile $\rho \propto r^{-3.5}$ with a spherical shape.

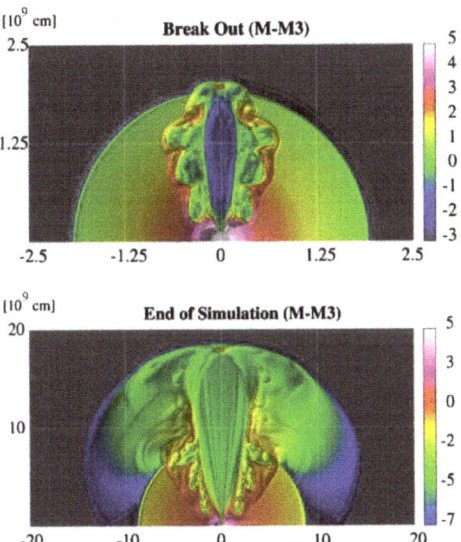

Figure 9. Two snapshots from a model of [192]. The top panel is at the time where the jet breaks out and the lower panel at the end of the simulation. The jet was injected at 50 ms after merger with an opening angle of 15°. The average opening angle of the jet after break out is 12.6°. (Reprinted from [192]. © AAS. Reproduced with permission).

In [193], they studied the influence of the neutrino driven wind on the expansion and propagation of the formed jet, considering the post-merger production of neutrino fluxes that contribute to a wind density profile. They quantified this wind as [90,196,197]:

$$\dot{M}_w \sim 5 \times 10^{-4} \left(\frac{L_\nu}{10^{52} \text{ erg s}^{-1}} \right) M_\odot \text{ s}^{-1}, \tag{11}$$

which results in a limiting Lorentz factor for the jet as:

$$\Gamma_\nu \sim 10 \left(\frac{L_{jet}}{10^{52} \text{ erg s}^{-1}} \right) \left(\frac{\dot{M}_w}{5 \times 10^{-4} M_\odot \text{ s}^{-1}} \right). \tag{12}$$

Their wind profile depends on how long the neutrino driven wind was active. At the time that the wind stops, a jet is injected. In Figure 10 (left panel), a parameter study is presented on whether the jet can break out or not from such a wind. The axes are the luminosity of the jet versus time, where t_w depicts the time that the neutrino wind stops, supposedly when the merger remnant collapses to a BH. Matter is injected in the wind as $\dot{M}_w \sim 10^{-3} M_\odot \text{ s}^{-1}$ with a velocity of $u \sim 0.3c$. The coloured lines indicate a different termination time for the neutrino wind, where t_w is the time that the neutrino wind stops. As a comparisonm the T_{90} distribution (the duration distribution of short GRBs from [78,198]) is overplotted to show that when the neutrino wind operates for more than $t_w > 0.1$ s then jet duration times that exceed the observed ones are needed. Interestingly, all jets with luminosities less than 10^{51} erg s^{-1} are choked and never break out from the neutrino wind. This can be regarded as the limiting value for the production of a structured jet with a cocoon or a choked jet with a cocoon.

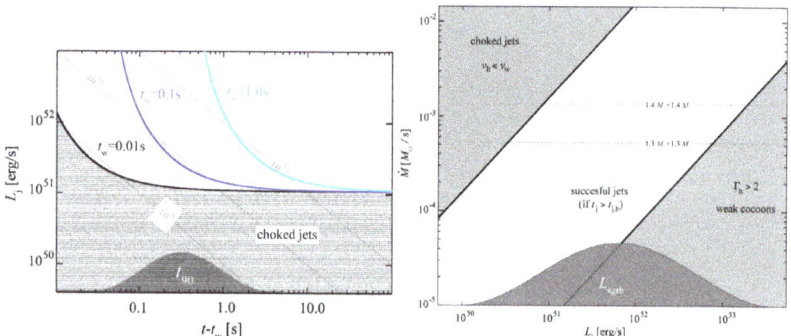

Figure 10. The left panel shows luminosity versus time, where t_w is the lifetime of the neutrino wind and the time that the jet begins to expand. The coloured lines indicate a model from [193], that has a wind injection rate of $\dot{M}_w \sim 10^{-3} M_\odot \text{ s}^{-1}$ with a velocity of 0.3c. Each line indicate a different termination time $t_w \sim 0.01, 0.1, 1$ s. For such a heavy wind, the luminosity of the jet has to be above 10^{51} erg s^{-1} and operate for at least the same time as the wind. (Reprinted from [193]. © AAS. Reproduced with permission).

However, this result strongly depends on the amount of mass that is ejected through this process. Thus, the next thing to compare is jet luminosity with respect to the mass injection from the neutrino wind. This result is shown in Figure 10 (right panel). The mass injection rate is plotted versus the jet luminosity and depicts different regions in the parameter space. If the luminosity is low (on the left part of the figure), then the velocity of the head of the jet is not exceeding the velocity of the wind and consequently never breaks out, resulting in a choked jet with a cocoon. Even for smaller luminosities, if the mass injection is less than 10^{-3}–$10^{-4} M_\odot \text{ s}^{-1}$, then a successful jet can be formed. It is also known that in order to produce a successful jet, the jet injection time has to exceed the break out time through any medium. They further comment on the production of a cocoon as the jet advances through the ejecta and deposits some of its energy to form such a cocoon [186].

Spherical versus oblate BNS ejecta. In the previously mentioned studies, the shape of the density profile that is mimicking the BNS ejecta was spherical. Thus, all results have to be interpreted as arising from within a spherically expanding mass cloud. However, there is a possibility that this is not true [199]. Recent simulations of BNS mergers indeed show that the merger ejecta and/or the post-merger-driven winds are not at all spherical [101–103,105,106,108,120]. In [200], they considered the interaction of the jet with an oblate mass cloud mimicking the BNS ejecta, as opposed to a spherical one. The earlier idea that the ejecta can provide the collimation of the jet [201] is stronger in the case where the BNS ejecta have an elongated shape. They inject a luminosity of:

$$L = eM_{cloud}c^2/\tau, \qquad(13)$$

where M_{cloud} is the mass of the cloud, e is the ratio of the energy deposited in the mass cloud and τ is the engine duration. Engines that act through an oblate cloud can collimate even wider initial angles. When the overall injected energy from the injected luminosity (Equation (13)) is low, then the kinetic energy of the dynamical ejecta can be higher and this does not allow for collimation. In the other limit where the injected energy is large, then the mass of the ejecta cannot provide sufficient collimation of the outflow. In the latter case, the outflow maintains the initial opening angle. For an initial opening angle of 29°, with a mass cloud of 10^{-4} $M_\odot$ and oblate in shape, a jet with luminosity of 10^{48}–10^{49} erg s^{-1} is significantly collimated with a resulting opening angle of 5–8° when breaking out from the cloud. In Figure 11, a model is shown from [200]. In this model, the mass cloud that mimics the BNS ejecta has an oblate shape. It is clearly seen that the interaction through the oblate mass cloud produces a narrow outflow with high Lorentz factor. We should also note here the possibility that jet formation may also account for the production of a magnetar after the BNS merger [202,203].

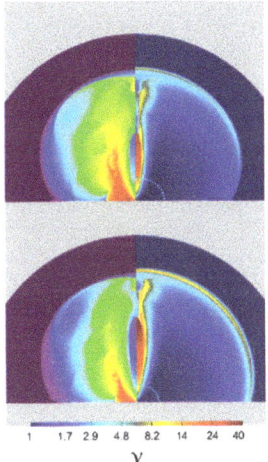

Figure 11. A model from [200] where the shape of the BNS ejecta is assumed to be oblate. In the upper and lower rows, two different times are depicted: $ct/R_0 = 15, 25$, where $R_0 = 850$ km is the initial radius of the mass cloud. The initial opening angle of both models is 60°. The left panel of each plot shows the density and the right panel of each plot shows the Lorentz factor. The outer surface of the expanding mass ejecta is depicted with a dashed cyan curve. The model depicted in this figure with the oblate shaped cloud clearly produces a narrow relativistic outflow. In both cases, the ratio between the energy of the engine to the rest mass energy of the ejecta is $E_{engine}/M_0c^2 = 0.024$, where $M_0 = 10^{-4}$ $M_\odot$ is the mass of the cloud ejecta. (Reprinted from [200]. © AAS. Reproduced with permission).

The next step was to use more realistic profiles taken from [98,204] to continue a more detailed study of the interaction of the jet with the neutrino-driven and magnetically-driven wind, as studied in [194]. They concluded that a jet with luminosity comparable to the observed ones from short GRBs can break out from such winds with the requirement of having an initial opening angle of $\lesssim 20°$. They further used the observed duration of short GRBs to set limits on the lifetime of the production of winds from a HMNS, which is determined by the time that the jet needs to break out.

Observables from off-axis emission. All such simulations act as a first step towards understanding the jet and cocoon observables that follow a BNS merger. The next step was to see how these components would show up when observed off-axis. Furthermore, late radio counterparts from BNS mergers have long been proposed and expected [205]. Wide angle signatures from jet and cocoon interactions were presented through semi-analytical calculations in [206]. They calculated the on-axis and off-axis emission of a short GRB. They included the prompt and afterglow emission from a relativistic jet, as well as the prompt and afterglow emission from the cocoon formed through the interaction of the jet and the surrounding ejected material. The energy of the cocoon was found to amount to approximately 10% of the energy of the burst itself. However, the cocoon energy strongly depends on the structure and size of the ejected material.

In the case of long GRBs, the propagation of the jet through a baryon loaded region (such as the interior of a massive star) has been studied and provides a clear and robust observational picture. Nakar and Piran [207] made a comprehensive (mostly analytical) study of the observable signatures of GRB cocoons. Their main focus was on the collapsar model for long GRBs, which envisions the propagation of a jet inside a massive star. While their focus was on cocoons emerging from long GRBs, short GRB cocoons should have an analogous signature (maybe weaker) as they indicated. All the formulas and equations reported in this study can provide a quick in-depth description of the characteristics of a cocoon and its emission. The analytical modelling in [208], calibrated by numerical results from [191], can be used to estimate the cocoon parameters through the jet break out time and the characteristics of the ejected matter.

Simulations of short GRB cocoons can provide more details on the production of the cocoon itself, together with realistic characteristics for its shape and initial Lorentz factor, which are key elements for a realistic description of any observables coming from it [209,210]. The numerical setup by Lazzati et al. [209] is an injected jet with luminosity of $L_j = 10^{50}$ erg s^{-1}, an initial opening angle of $\theta_j = 16°$ and the duration of this engine was defined to be $t_{enigne} = 1$ s.

Through the isotropic equivalent energy three different components can be identified. The core of the outflow, which is the initially injected jet modified through the interaction with the ejecta and is the brightest part confined in $\theta_{jet} \sim 15°$. The surrounding material of the jet that forms a hot bubble is the energized cocoon which occupies a region within 15–45°. The third component is a fairly isotropic wide-angle structure that stops at an angle of 65°. From the initial energy of 10^{50} erg that was injected, 5.5×10^{49} erg remains in the confined jet, 3.8×10^{48} erg are given to the surrounding cocoon and 7×10^{47} erg are found in the shocked ambient medium. The rest of the energy is stored in slow moving material ($\Gamma < 1.1$). Figure 12 shows the results from [209], where the left panel shows the isotropic equivalent energy where the three components can be identified, and the right panel the peak photon energy is plotted as seen from different angles. The cocoon emission was also studied in detail by Gottlieb et al. [210]. Their main focus was the appearance of a kilonova following the radioactive heating of the merger ejecta.

Jet with core and sheath. In a similar spirit, Kathirgamaraju et al. [211] simulated the off-axis emission from a short GRB jet including magnetic field. They argued that for a realistic jet model, one whose Lorentz factor and luminosity vary smoothly with angle, detection can be achieved for a broader range of viewing angles. In Figure 13, the luminosity and Lorentz factor is shown from their model. It is clear that even for angles larger than 20° the luminosity from the jet is significant. As the jet breaks out from the cocoon, the prompt emission is released [203,212]. The time that the shock breaks out is pictured from a simulation of [212], illustrated in their Figure 1. As the shock propagates through the expanding BNS ejecta it accumulates mass on top of the jet head. The wide parts of the jet are not collimated and they propagate conically inside the mass cloud. If the engine operates for long enough, the shock breaks out and it is not choked inside the ejecta after giving all its energy to them. The break out of this shock in the magetized case was studied by [203].

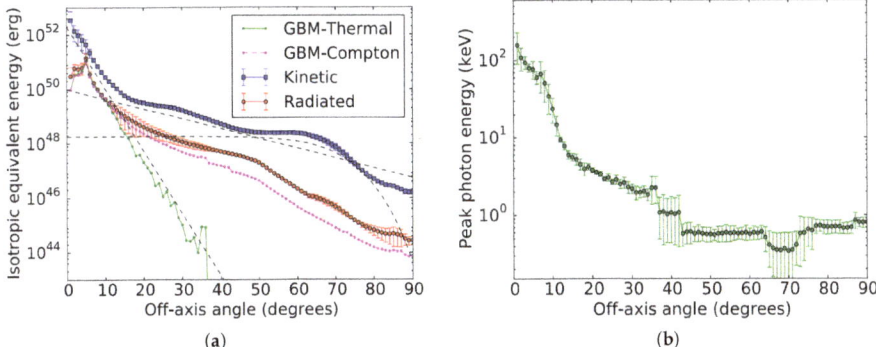

Figure 12. Both figures are taken from [209]. (**a**) Off-axis distribution of the isotropic equivalent energy. The error bars show the range of variation at each specific angle. The kinetic energy is shown in blue squares, while the bolometric energy is shown in red dots. The energy that Fermi (Gamma-ray Burst Monitor (GBM)) would detect is shown as lines with dots, green solid line is for a thermal spectrum and magenta dashed line for a Comptonized spectrum. The three components: the jet (exponential), the cocoon (exponential), and the shocked ambient medium (constant with sharp cutoff) are overlaid on the kinetic energy profile as black dashed lines. (**b**) Off-axis emission from the jet/cocoon photosphere. The peak photon energy is depicted, while the symbols are as defined above. (Reprinted from [209]. © AAS. Reproduced with permission).

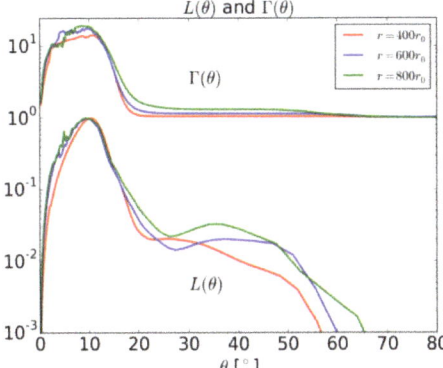

Figure 13. The appearance of a jet model after break out from the BNS ejecta. A model from [211]. This figure presents the jet luminosity $\Lambda(\theta)$ (in arbitrary units) and Lorentz factor as a function of the observer's angle. Quantities are extracted at three different radii. It is evident that even for angles greater than 20°, the luminosity is reduced but still significant. (Reprinted from [211]. © Oxford University Press. Reproduced with permission).

Magnetic explosion. It was argued in the last part of Section 2 that if the collapse of the compact remnant comes late (after a second), then the small amount of mass left at the torus cannot give a sufficient boundary for a magnetic jet to be launched. As such, all the magnetic energy dissipates away and produces an explosion. In Figure 14, such an explosion is depicted at the time that the outflow has entered a low density region. The main characteristics of a cocoon are still entering this picture.

A main difference is that there does not exist an easily distinguishable relativistic core with a small opening angle. Faster moving material can be found at larger angles, as can be seen in Figure 14. This is a model from an upcoming work.

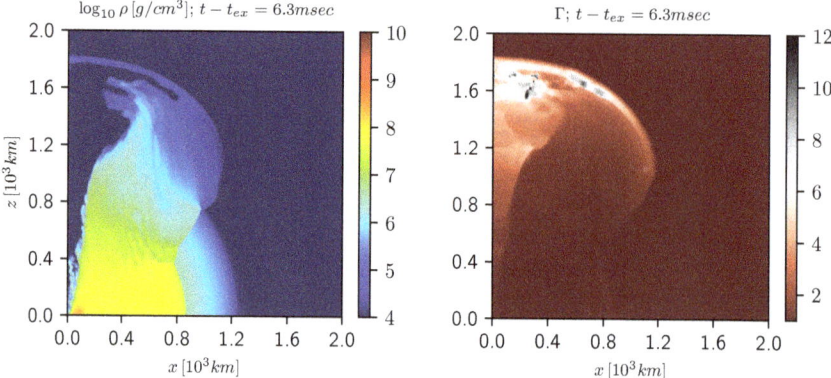

Figure 14. A late magnetic explosion triggered by the collapse of the compact remnant is shown, a model similar to the one from an upcoming work. The amount of magnetic energy released is on the order of 5×10^{51} erg. The left panel shows the density and the right panel the Lorentz factor. The snapshot is taken at the time that the shock enters the low density region and expands sideways. It is interesting to note that there is no clear relativistic core with a small opening angle, although there are regions of the outflow at larger angles that move slightly faster.

Afterglow. In late observations, following a BNS merger event, it is important to understand the signatures from different components and the differences in observations from different models. As the outflow that was produced from the BNS merger hits the ISM, a shock is produced wherein particles are energized and emit synchrotron radiation. The outflow continues to sweep up matter and begin to decelerate. This is the standard picture for the source of a GRB afterglow. In the case of short GRBs from BNS mergers this has been discussed significantly before the detection of GW170817 [205,213,214]. Afterglow model predictions from numerical simulations have been studied in the context of long GRBs (e.g., [215]), and also as seen off-axis [216]. After the coincident detection of GW170817 together with GRB170817A and the following afterglow observations, there is an enormous effort to analyze the data and fit them with realistic models in order to clarify what are the actual components that powered such emission. It would be unrealistic to review such ongoing efforts. We restrict ourselves to a brief overview of observations and the corresponding modelling of them.

The prompt gamma-ray emission was reported in [3,4]. The first detection of X-rays from the event came nine days later [49,50], whereas the first radio observations came sixteen days after merger [51]. The first interpretation acknowledged that we are observing something quite different to other short GRBs [14,54].

Ongoing efforts in understanding the EM counterparts of GW170817 include: afterglow modelng through hydrodynamic simulations of a jet propagating through the merger ejecta [217–219], radio imaging that could show the exact morphology of the outflow and polarization measurements that could help to distinguish different outflow structures [220–223]. Ideas that the merger event did not include a jet have been proposed [224–226] or models that follow the canonical picture with a short GRB jet [227,228]. Observation of GW170817 can provide a deep understanding of short GRB modelling [184,229,230]. It has also been proposed that the afterglow may come from the interaction of the fast tail of the BNS ejecta with the ISM [231]. Another indicator may be the appearance of the counter jet [232], and how to probe short GRB properties from GW events [233]. Furthermore, one may also ask how the magnetar model can be in the picture [234].

Before finishing this section, we would like to gather some important points that should be kept in mind for the study of a relativistic outflow passing through the BNS ejecta. **Critical points:**

- The amount of dynamically ejected matter strongly depends on the total mass of the binary and the mass ratio [88–108].
- The BNS ejecta are not spherical, rather they have a unique structure for every different EOS used [106].
- The time that the engine begins to produce an outflow is extremely important, since this will depict how much mass has been ejected by neutrino and magnetic winds [98,194,204].

4. Conclusions

In the years to come, more detections of BNS mergers are expected from ground-based interferometers. Combined observations of GW and EM radiation of such events would have a great impact on theoretical and numerical studies discussed here. The theoretical modeling of such events will have enormous benefit from the observational signatures of BNS mergers expected to be gathered in the next years. Our understanding of such extreme events lies in the comparison of these theoretical models against observations. It is important to analyze in detail observations of GW170817 and all its EM counterparts, starting with GRB170817A. It is equally important to reproduce realistic physics through numerical simulations to match and explain observations. This brief review can act as a quick introduction to BNS numerical relativity simulations for people interested in short GRB outflows through BNS ejecta, or as a brief introduction to short GRB jet simulations and setups used by people working on BNS merger simulations. Overall, we want to point out the importance of combining knowledge from both paths in order for a consistent picture to be drawn at the end.

In Section 2, we went through studies from numerical relativity for magnetized BNS mergers. We highlight important aspects of this physical process as given in the literature. Issues, such as the magnetic field amplification, the difficulty of launching a relativistic jet, the mass ejection during merger, and all possible winds produced after merger, can become clear through detailed studies. At the end of the section, we state several important points (importance is a subjective criterion).

The next step is to take these different ingredients from BNS simulations and study any outflow emerging after merger. A relativistic outflow has been observed from a BNS merger [52,53]. Thus, we need to understand how it was launched, what is the initial structure of this outflow, and how it will evolve through its interaction with the BNS ejecta. In Section 3, we briefly go through previous works on these aspects. This is a rapidly evolving sub-field, especially after the detection. Now, any model and idea can be simulated and be exposed to the data that followed GW170817. However, we should keep in mind that a BNS can have a different evolution, even with a very small difference in mass. In the end, modelling and studying outflows of such events should be inspired by GW170817.

Funding: This research was funded by [Alexander von Humboldt-Stiftung].

Acknowledgments: It is a pleasure to thank D. Giannios, R. Gold, E. Most and V. Paschalidis for reading the manuscript and giving valuable comments. The author is supported by an Alexander von Humboldt Fellowship.

Conflicts of Interest: The authors declare no conflict of interest.

References

1. The LIGO Scientific Collaboration; The Virgo Collaboration. GW170817: Observation of Gravitational Waves from a Binary Neutron Star Inspiral. *Phys. Rev. Lett.* **2017**, *119*, 161101. [CrossRef] [PubMed]
2. The LIGO Scientific Collaboration; The Virgo Collaboration; Abbott, B.P.; Abbott, R.; Abbott, T.D.; Acernese, F.; Ackley, K.; Adams, C.; Adams, T.; Addesso, P.; et al. Multi-messenger Observations of a Binary Neutron Star Merger. *Astrophys. J. Lett.* **2017**, *848*, L12. [CrossRef]
3. Goldstein, A.; Veres, P.; Burns, E.; Briggs, M.S.; Hamburg, R.; Kocevski, D.; Wilson-Hodge, C.A.; Preece, R.D.; Poolakkil, S.; Roberts, O.J.; et al. An Ordinary Short Gamma-Ray Burst with Extraordinary Implications: Fermi-GBM Detection of GRB 170817A. *Astrophys. J. Lett.* **2017**, *848*, L14. [CrossRef]

4. Savchenko, V.; Ferrigno, C.; Kuulkers, E.; Bazzano, A.; Bozzo, E.; Brandt, S.; Chenevez, J.; Courvoisier, T.J.L.; Diehl, R.; Domingo, A.; et al. INTEGRAL Detection of the First Prompt Gamma-Ray Signal Coincident with the Gravitational-wave Event GW170817. *Astrophys. J. Lett.* **2017**, *848*, L15. [CrossRef]
5. Coulter, D.A.; Foley, R.J.; Kilpatrick, C.D.; Drout, M.R.; Piro, A.L.; Shappee, B.J.; Siebert, M.R.; Simon, J.D.; Ulloa, N.; Kasen, D.; et al. Swope Supernova Survey 2017a (SSS17a), the optical counterpart to a gravitational wave source. *Science* **2017**, *358*, 1556–1558. [CrossRef] [PubMed]
6. Soares-Santos, M.; Holz, D.E.; Annis, J.; Chornock, R.; Herner, K.; Berger, E.; Brout, D.; Chen, H.Y.; Kessler, R.; Sako, M.; et al. The Electromagnetic Counterpart of the Binary Neutron Star Merger LIGO/Virgo GW170817. I. Discovery of the Optical Counterpart Using the Dark Energy Camera. *Astrophys. J. Lett.* **2017**, *848*, L16. [CrossRef]
7. Arcavi, I.; Hosseinzadeh, G.; Howell, D.A.; McCully, C.; Poznanski, D.; Kasen, D.; Barnes, J.; Zaltzman, M.; Vasylyev, S.; Maoz, D.; et al. Optical emission from a kilonova following a gravitational-wave-detected neutron-star merger. *Nature* **2017**, *551*, 64–66. [CrossRef]
8. Nicholl, M.; Berger, E.; Kasen, D.; Metzger, B.D.; Elias, J.; Briceño, C.; Alexander, K.D.; Blanchard, P.K.; Chornock, R.; Cowperthwaite, P.S.; et al. The Electromagnetic Counterpart of the Binary Neutron Star Merger LIGO/Virgo GW170817. III. Optical and UV Spectra of a Blue Kilonova from Fast Polar Ejecta. *Astrophys. J. Lett.* **2017**, *848*, L18. [CrossRef]
9. Pian, E.; D'Avanzo, P.; Benetti, S.; Branchesi, M.; Brocato, E.; Campana, S.; Cappellaro, E.; Covino, S.; D'Elia, V.; Fynbo, J.P.U.; et al. Spectroscopic identification of r-process nucleosynthesis in a double neutron-star merger. *Nature* **2017**, *551*, 67–70. [CrossRef] [PubMed]
10. Smartt, S.; Chen, T.-W.; Jerkstrand, A.; Coughlin, M.; Kankare, E.; Sim, S.A.; Fraser, M.; Inserra, C.; Maguire, K.; Chambers, K.C.; et al. A kilonova as the electromagnetic counterpart to a gravitational-wave source. *Nature* **2017**, *551*, 75–79. [CrossRef] [PubMed]
11. Tanvir, N.R.; Levan, A.J.; González-Fernández, C.; Korobkin, O.; Mandel, I.; Rosswog, S.; Hjorth, J.; D'Avanzo, P.; Fruchter, A.S.; Fryer, C.L.; et al. The Emergence of a Lanthanide-rich Kilonova Following the Merger of Two Neutron Stars. *Astrophys. J. Lett.* **2017**, *848*, L27. [CrossRef]
12. Utsumi, Y.; Tanaka, M.; Tominaga, N.; Yoshida, M.; Barway, S.; Nagayama, T.; Zenko, T.; Aoki, K.; Fujiyoshi, T.; Furusawa, H.; et al. J-GEM observations of an electromagnetic counterpart to the neutron star merger GW170817. *Publ. Astr. Soc. Jpn.* **2017**, *69*, 101. [CrossRef]
13. Kilpatrick, C.D.; Foley, R.J.; Kasen, D.; Murguia-Berthier, A.; Ramirez-Ruiz, E.; Coulter, D.A.; Drout, M.R.; Piro, A.L.; Shappee, B.J.; Boutsia, K.; et al. Electromagnetic evidence that SSS17a is the result of a binary neutron star merger. *Science* **2017**, *358*, 1583–1587. [CrossRef] [PubMed]
14. Kasliwal, M.M.; Nakar, E.; Singer, L.P.; Kaplan, D.L.; Cook, D.O.; Van Sistine, A.; Lau, R.M.; Fremling, C.; Gottlieb, O.; Jencson, J.E.; et al. Illuminating gravitational waves: A concordant picture of photons from a neutron star merger. *Science* **2017**, *358*, 1559–1565. [CrossRef] [PubMed]
15. Covino, S.; Wiersema, K.; Fan, Y.Z.; Toma, K.; Higgins, A.B.; Melandri, A.; D'Avanzo, P.; Mundell, C.G.; Wijers, R.A.M.J. The unpolarized macronova associated with the gravitational wave event GW 170817. *Nat. Astron.* **2017**, *1*, 791–794. [CrossRef]
16. Cowperthwaite, P.S.; Berger, E.; Villar, V.A.; Metzger, B.D.; Nicholl, M.; Chornock, R.; Blanchard, P.K.; Fong, W.; Margutti, R.; Soares-Santos, M.; et al. The Electromagnetic Counterpart of the Binary Neutron Star Merger LIGO/Virgo GW170817. II. UV, Optical, and Near-infrared Light Curves and Comparison to Kilonova Models. *Astrophys. J. Lett.* **2017**, *848*, L17. [CrossRef]
17. Buckley, D.A.H.; Andreoni, I.; Barway, S.; Cooke, J.; Crawford, S.M.; Gorbovskoy, E.; Gromadzki, M.; Lipunov, V.; Mao, J.; Potter, S.B.; et al. A comparison between SALT/SAAO observations and kilonova models for AT 2017gfo: The first electromagnetic counterpart of a gravitational wave transient—GW170817. *Mon. Not. R. Astron. Soc.* **2018**, *474*, L71–L75. [CrossRef]
18. Drout, M.R.; Piro, A.L.; Shappee, B.J.; Kilpatrick, C.D.; Simon, J.D.; Contreras, C.; Coulter, D.A.; Foley, R.J.; Siebert, M.R.; Morrell, N.; et al. Light Curves of the Neutron Star Merger GW170817/SSS17a: Implications for R-Process Nucleosynthesis. *Science* **2017**, *358*, 1570–1574. [CrossRef] [PubMed]
19. Evans, P.A.; Cenko, S.B.; Kennea, J.A.; Emery, S.W.K.; Kuin, N.P.M.; Korobkin, O.; Wollaeger, R.T.; Tagliaferri, G.; Tanvir, N.R.; Tohuvavohu, A. Swift and NuSTAR observations of GW170817: Detection of a blue kilonova. *Science* **2017**, *358*, 1565–1570. [CrossRef] [PubMed]

20. Arcavi, I. The First Hours of the GW170817 Kilonova and the Importance of Early Optical and Ultraviolet Observations for Constraining Emission Models. *Astrophys. J. Lett.* **2018**, *855*, L23. [CrossRef]
21. Valenti, S.; David.; Sand, J.; Yang, S.; Cappellaro, E.; Tartaglia, L.; Corsi, A.; Jha, S.W.; Reichart, D.E.; Haislip, J.; et al. The Discovery of the Electromagnetic Counterpart of GW170817: Kilonova AT 2017gfo/DLT17ck. *Astrophys. J. Lett.* **2017**, *848*, L24. [CrossRef]
22. Eichler, D.; Livio, M.; Piran, T.; Schramm, D.N. Nucleosynthesis, neutrino bursts and gamma-rays from coalescing neutron stars. *Nature* **1989**, *340*, 126–128. [CrossRef]
23. Narayan, R.; Paczynski, B.; Piran, T. Gamma-ray bursts as the death throes of massive binary stars. *Astrophys. J. Lett.* **1992**, *395*, L83–L86. [CrossRef]
24. Mochkovitch, R.; Hernanz, M.; Isern, J.; Martin, X. Gamma-ray bursts as collimated jets from neutron star/black hole mergers. *Nature* **1993**, *361*, 236. [CrossRef]
25. Annala, E.; Gorda, T.; Kurkela, A.; Vuorinen, A. Gravitational-Wave Constraints on the Neutron-Star-Matter Equation of State. *Phys. Rev. Lett.* **2018**, *120*, 172703. [CrossRef] [PubMed]
26. Bauswein, A.; Just, O.; Janka, H.T.; Stergioulas, N. Neutron-star Radius Constraints from GW170817 and Future Detections. *Astrophys. J. Lett.* **2017**, *850*, L34. [CrossRef]
27. Margalit, B.; Metzger, B.D. Constraining the Maximum Mass of Neutron Stars from Multi-messenger Observations of GW170817. *Astrophys. J. Lett.* **2017**, *850*, L19. [CrossRef]
28. Radice, D.; Perego, A.; Zappa, F.; Bernuzzi, S. GW170817: Joint Constraint on the Neutron Star Equation of State from Multimessenger Observations. *Astrophys. J. Lett.* **2018**, *852*, L29. [CrossRef]
29. Rezzolla, L.; Most, E.R.; Weih, L.R. Using Gravitational-wave Observations and Quasi-universal Relations to Constrain the Maximum Mass of Neutron Stars. *Astrophys. J. Lett.* **2018**, *852*, L25. [CrossRef]
30. Ruiz, M.; Shapiro, S.L.; Tsokaros, A. GW170817, general relativistic magnetohydrodynamic simulations, and the neutron star maximum mass. *Phys. Rev. D* **2018**, *97*, 021501. [CrossRef] [PubMed]
31. Shibata, M.; Fujibayashi, S.; Hotokezaka, K.; Kiuchi, K.; Kyutoku, K.; Sekiguchi, Y.; Tanaka, M. Modeling GW170817 based on numerical relativity and its implications. *Phys. Rev. D* **2017**, *96*, 123012. [CrossRef]
32. De, S.; Finstad, D.; Lattimer, J.M.; Brown, D.A.; Berger, E.; Biwer, C.M. Constraining the nuclear equation of state with GW170817. *arXiv* **2018**, arXiv:1804.08583.
33. Tews, I.; Carlson, J.; Gandolfi, S.; Reddy, S. Constraining the speed of sound inside neutron stars with chiral effective field theory interactions and observations. *Astrophys. J.* **2018**, *860*, 149. [CrossRef]
34. Tews, I.; Margueron, J.; Reddy, S. How well does GW170817 constrain the equation of state of dense matter? *arXiv* **2018**, arXiv:1804.02783.
35. Alsing, J.; Silva, H.O.; Berti, E. Evidence for a maximum mass cut-off in the neutron star mass distribution and constraints on the equation of state. *Mon. Not. R. Astron. Soc.* **2018**, *478*, 1377–1391. [CrossRef]
36. Burgio, G.F.; Drago, A.; Pagliara, G.; Schulze, H.J.; Wei, J.B. Has deconfined quark matter been detected during GW170817/AT2017gfo? *arXiv* **2018**, arXiv:1803.09696.
37. Raithel, C.; Özel, F.; Psaltis, D. Tidal deformability from GW170817 as a direct probe of the neutron star radius. *arXiv* **2018**, arXiv:1803.07687.
38. Paschalidis, V.; Yagi, K.; Alvarez-Castillo, D.; Blaschke, D.B.; Sedrakian, A. Implications from GW170817 and I-Love-Q relations for relativistic hybrid stars. *Phys. Rev. D* **2018**, *97*, 084038. [CrossRef]
39. Lim, Y.; Holt, J.W. Neutron Star Tidal Deformabilities Constrained by Nuclear Theory and Experiment. *Phys. Rev. Lett.* **2018**, *121*, 062701. [CrossRef] [PubMed]
40. Lattimer, J.M.; Schramm, D.N. Black-hole-neutron-star collisions. *Astrophys. J. Lett.* **1974**, *192*, L145–L147. [CrossRef]
41. Li, L.X.; Paczynski, B. Transient events from neutron star mergers. *Astrophys. J.* **1998**, *507*, L59. [CrossRef]
42. Kasen, D.; Metzger, B.; Barnes, J.; Quataert, E.; Ramirez-Ruiz, E. Origin of the heavy elements in binary neutron-star mergers from a gravitational-wave event. *Nature* **2017**, *551*, 80–84. [CrossRef] [PubMed]
43. Tanaka, M.; Utsumi, Y.; Mazzali, P.A.; Tominaga, N.; Yoshida, M.; Sekiguchi, Y.; Morokuma, T.; Motohara, K.; Ohta, K.; Kawabata, K.S.; et al. Kilonova from post-merger ejecta as an optical and near-Infrared counterpart of GW170817. *Publ. Astron. Soc. Jpn.* **2017**, *69*, 102. [CrossRef]
44. Murguia-Berthier, A.; Ramirez-Ruiz, E.; Kilpatrick, C.D.; Foley, R.J.; Kasen, D.; Lee, W.H.; Piro, A.L.; Coulter, D.A.; Drout, M.R.; Madore, B.F.; et al. A Neutron Star Binary Merger Model for GW170817/GRB 170817A/SSS17a. *Astrophys. J. Lett.* **2017**, *848*, L34. [CrossRef]

45. Waxman, E.; Ofek, E.; Kushnir, D.; Gal-Yam, A. Constraints on the ejecta of the GW170817 neutron-star merger from its electromagnetic emission. *arXiv* 2017, arXiv:1711.09638.
46. Villar, V.A.; Guillochon, J.; Berger, E.; Metzger, B.D.; Cowperthwaite, P.S.; Nicholl, M.; Alexander, K.D.; Blanchard, P.K.; Chornock, R.; Eftekhari, T.; et al. The Combined Ultraviolet, Optical, and Near-infrared Light Curves of the Kilonova Associated with the Binary Neutron Star Merger GW170817: Unified Data Set, Analytic Models, and Physical Implications. *Astrophys. J. Lett.* 2017, *851*, L21. [CrossRef]
47. Perego, A.; Radice, D.; Bernuzzi, S. AT 2017gfo: An Anisotropic and Three-component Kilonova Counterpart of GW170817. *Astrophys. J. Lett.* 2017, *850*, L37. [CrossRef]
48. Metzger, B.D.; Thompson, T.A.; Quataert, E. A Magnetar Origin for the Kilonova Ejecta in GW170817. *Astrophys. J.* 2018, *856*, 101. [CrossRef]
49. Troja, E.; Piro, L.; van Eerten, H.; Wollaeger, R.T.; Im, M.; Fox, O.D.; Butler, N.R.; Cenko, S.B.; Sakamoto, T.; Fryer, C.L.; et al. The X-ray counterpart to the gravitational-wave event GW170817. *Nature* 2017, *551*, 71–74. [CrossRef]
50. Margutti, R.; Berger, E.; Fong, W.; Guidorzi, C.; Alexander, K.D.; Metzger, B.D.; Blanchard, P.K.; Cowperthwaite, P.S.; Chornock, R.; Eftekhari, T.; et al. The Electromagnetic Counterpart of the Binary Neutron Star Merger LIGO/Virgo GW170817. V. Rising X-ray Emission from an Off-axis Jet. *Astrophys. J. Lett.* 2017, *848*, L20. [CrossRef]
51. Hallinan, G.; Corsi, A.; Mooley, K.P.; Hotokezaka, K.; Nakar, E.; Kasliwal, M.M.; Kaplan, D.L.; Frail, D.A.; Myers, S.T.; Murphy, T.; et al. A radio counterpart to a neutron star merger. *Science* 2017, *358*, 1579–1583. [CrossRef] [PubMed]
52. Alexander, K.D.; Berger, E.; Fong, W.; Williams, P.K.G.; Guidorzi, C.; Margutti, R.; Metzger, B.D.; Annis, J.; Blanchard, P.K.; Brout, D.; et al. The Electromagnetic Counterpart of the Binary Neutron Star Merger LIGO/Virgo GW170817. VI. Radio Constraints on a Relativistic Jet and Predictions for Late-time Emission from the Kilonova Ejecta. *Astrophys. J. Lett.* 2017, *848*, L21. [CrossRef]
53. Haggard, D.; Nynka, M.; Ruan, J.J.; Kalogera, V.; Cenko, S.B.; Evans, P.; Kennea, J.A. A Deep Chandra X-ray Study of Neutron Star Coalescence GW170817. *Astrophys. J. Lett.* 2017, *848*, L25. [CrossRef]
54. Granot, J.; Guetta, D.; Gill, R. Lessons from the Short GRB 170817A: The First Gravitational-wave Detection of a Binary Neutron Star Merger. *Astrophys. J. Lett.* 2017, *850*, L24. [CrossRef]
55. Mooley, K.P.; Nakar, E.; Hotokezaka, K.; Hallinan, G.; Corsi, A.; Frail, D.A.; Horesh, A.; Murphy, T.; Lenc, E.; Kaplan, D.L.; et al. A mildly relativistic wide-angle outflow in the neutron-star merger event GW170817. *Nature* 2018, *554*, 207–210. [CrossRef] [PubMed]
56. Ruan, J.J.; Nynka, M.; Haggard, D.; Kalogera, V.; Evans, P. Brightening X-ray Emission from GW170817/GRB 170817A: Further Evidence for an Outflow. *Astrophys. J. Lett.* 2018, *853*, L4. [CrossRef]
57. Pooley, D.; Kumar, P.; Wheeler, J.C.; Grossan, B. GW170817 Most Likely Made a Black Hole. *Astrophys. J. Lett.* 2018, *859*, L23. [CrossRef]
58. Margutti, R.; Alexander, K.D.; Xie, X.; Sironi, L.; Metzger, B.D.; Kathirgamaraju, A.; Fong, W.; Blanchard, P.K.; Berger, E.; MacFadyen, A.; et al. The Binary Neutron Star Event LIGO/Virgo GW170817 160 Days after Merger: Synchrotron Emission across the Electromagnetic Spectrum. *Astrophys. J. Lett.* 2018, *856*, L18. [CrossRef]
59. Lyman, J.D.; Lamb, G.P.; Levan, A.J.; Mandel, I.; Tanvir, N.R.; Kobayashi, S.; Gompertz, B.; Hjorth, J.; Fruchter, A.S.; Kangas, T.; et al. The optical afterglow of the short gamma-ray burst associated with GW170817. *Nat. Astron.* 2018, *2*, 751–754. [CrossRef]
60. Li, B.; Li, L.B.; Huang, Y.F.; Geng, J.J.; Yu, Y.B.; Song, L.M. Continued Brightening of the Afterglow of GW170817/GRB 170817A as Being Due to a Delayed Energy Injection. *Astrophys. J. Lett.* 2018, *859*, L3. [CrossRef]
61. Geng, J.J.; Dai, Z.G.; Huang, Y.F.; Wu, X.F.; Li, L.B.; Li, B.; Meng, Y.Z. Brightening X-ray/Optical/Radio Emission of GW170817/SGRB 170817A: Evidence for an Electron-Positron Wind from the Central Engine? *Astrophys. J. Lett.* 2018, *856*, L33. [CrossRef]
62. Dobie, D.; Kaplan, D.L.; Murphy, T.; Lenc, E.; Mooley, K.P.; Lynch, C.; Corsi, A.; Frail, D.; Kasliwal, M.; Hallinan, G. A Turnover in the Radio Light Curve of GW170817. *Astrophys. J. Lett.* 2018, *858*, L15. [CrossRef]
63. Alexander, K.D.; Margutti, R.; Blanchard, P.K.; Fong, W.; Berger, E.; Hajela, A.; Eftekhari, T.; Chornock, R.; Cowperthwaite, P.S.; Giannios, D.; et al. A Decline in the X-ray through Radio Emission from GW170817 Continues to Support an Off-Axis Structured Jet. *arXiv* 2018, arXiv:1805.02870.

64. Nynka, M.; Ruan, J.J.; Haggard, D.; Evans, P.A. Fading of the X-ray Afterglow of Neutron Star Merger GW170817/GRB170817A at 260 days. *arXiv* **2018**, arXiv:1805.04093.
65. Kruckow, M.U.; Tauris, T.M.; Langer, N.; Kramer, M.; Izzard, R.G. Progenitors of gravitational wave mergers: Binary evolution with the stellar grid-based code COMBINE. *Mon. Not. R. Astron. Soc.* **2018**, *481*, 1908–1949. [CrossRef]
66. Faber, J.A.; Rasio, F.A. Binary Neutron Star Mergers. *Living Rev. Relativ.* **2012**, *15*. [CrossRef] [PubMed]
67. Baiotti, L.; Rezzolla, L. Binary neutron-star mergers: A review of Einstein's richest laboratory. *Rep. Prog. Phys.* **2017**, *80*, 096901. [CrossRef] [PubMed]
68. Shibata, M.; Liu, Y.T.; Shapiro, S.L.; Stephens, B.C. Magnetorotational collapse of massive stellar cores to neutron stars: Simulations in full general relativity. *Phys. Rev. D* **2006**, *74*, 104026. [CrossRef]
69. Paschalidis, V. General relativistic simulations of compact binary mergers as engines for short gamma-ray bursts. *Class. Quantum Gravity* **2017**, *34*, 084002. [CrossRef]
70. Ciolfi, R. Short gamma-ray burst central engines. *arXiv* **2018**, arXiv:1804.03684.
71. Berger, E. Short-Duration Gamma-Ray Bursts. *Annu. Rev. Astron. Astrophys.* **2014**, *52*, 43–105. [CrossRef]
72. Fong, W.; Berger, E.; Margutti, R.; Zauderer, B.A. A Decade of Short-duration Gamma-Ray Burst Broadband Afterglows: Energetics, Circumburst Densities, and Jet Opening Angles. *Astrophys. J.* **2015**, *815*, 102. [CrossRef]
73. Rosswog, S. The multi-messenger picture of compact binary mergers. *Int. J. Mod. Phys. D* **2015**, *24*, 1530012. [CrossRef]
74. Fernández, R.; Quataert, E.; Schwab, J.; Kasen, D.; Rosswog, S. The interplay of disc wind and dynamical ejecta in the aftermath of neutron star-black hole mergers. *Mon. Not. R. Astron. Soc.* **2015**, *449*, 390–402. [CrossRef]
75. Thielemann, F.K.; Eichler, M.; Panov, I.V.; Wehmeyer, B. Neutron Star Mergers and Nucleosynthesis of Heavy Elements. *Annu. Rev. Nucl. Part. Sci.* **2017**, *67*, 253–274. [CrossRef]
76. Metzger, B.D. Kilonovae. *Living Rev. Relativ.* **2017**, *20*, 3. [CrossRef] [PubMed]
77. Paschalidis, V.; Stergioulas, N. Rotating stars in relativity. *Living Rev. Relativ.* **2017**, *20*, 7. [CrossRef] [PubMed]
78. Nakar, E. Short-hard gamma-ray bursts. *Phys. Rep.* **2007**, *442*, 166–236. [CrossRef]
79. Lee, W.H.; Ramirez-Ruiz, E. The Progenitors of Short Gamma-Ray Bursts. *New J. Phys.* **2007**, *9*, 17. [CrossRef]
80. Piran, T. The physics of gamma-ray bursts. *Rev. Mod. Phys.* **2005**, *76*, 1143–1210. [CrossRef]
81. Meszaros, P. Gamma-Ray Bursts. *Rep. Prog. Phys.* **2006**, *69*, 2259–2322. [CrossRef]
82. Kumar, P.; Smoot, G.F. Some implications of inverse-Compton scattering of hot cocoon radiation by relativistic jets in gamma-ray bursts. *Mon. Not. R. Astron. Soc.* **2014**, *445*, 528–543. [CrossRef]
83. Tauris, T.M.; Kramer, M.; Freire, P.C.C.; Wex, N.; Janka, H.T.; Langer, N.; Podsiadlowski, P.; Bozzo, E.; Chaty, S.; Kruckow, M.U.; et al. Formation of Double Neutron Star Systems. *Astrophys. J.* **2017**, *846*, 170. [CrossRef]
84. Baumgarte, T.W.; Shapiro, S.L.; Shibata, M. On the Maximum Mass of Differentially Rotating Neutron Stars. *Astrophys. J. Lett.* **2000**, *528*, L29–L32. [CrossRef]
85. Sekiguchi, Y.; Kiuchi, K.; Kyutoku, K.; Shibata, M. Gravitational Waves and Neutrino Emission from the Merger of Binary Neutron Stars. *Phys. Rev. Lett.* **2011**, *107*, 051102. [CrossRef] [PubMed]
86. Paschalidis, V.; Etienne, Z.B.; Shapiro, S.L. Importance of cooling in triggering the collapse of hypermassive neutron stars. *Phys. Rev. D* **2012**, *86*, 064032. [CrossRef]
87. Kaplan, J.D. Where Tori Fear to Tread: Hypermassive Neutron Star Remnants and Absolute Event Horizons or Topics in Computational General Relativity. Ph.D. Thesis, California Institute of Technology, Pasadena, CA, USA, 2014.
88. Rosswog, S.; Liebendörfer, M.; Thielemann, F.K.; Davies, M.B.; Benz, W.; Piran, T. Mass ejection in neutron star mergers. *Astron. Astrophys.* **1999**, *341*, 499–526.
89. Aloy, M.A.; Janka, H.; Müller, E. Relativistic outflows from remnants of compact object mergers and their viability for short gamma-ray bursts. *Astron. Astrophys.* **2005**, *436*, 273–311. [CrossRef]
90. Dessart, L.; Ott, C.D.; Burrows, A.; Rosswog, S.; Livne, E. Neutrino Signatures and the Neutrino-Driven Wind in Binary Neutron Star Mergers. *Astrophys. J.* **2009**, *690*, 1681–1705. [CrossRef]
91. Rezzolla, L.; Baiotti, L.; Giacomazzo, B.; Link, D.; Font, J.A. Accurate evolutions of unequal-mass neutron-star binaries: Properties of the torus and short GRB engines. *Class. Quantum Gravity* **2010**, *27*, 114105. [CrossRef]

92. Roberts, L.F.; Kasen, D.; Lee, W.H.; Ramirez-Ruiz, E. Electromagnetic Transients Powered by Nuclear Decay in the Tidal Tails of Coalescing Compact Binaries. *Astrophys. J. Lett.* **2011**, *736*, L21. [CrossRef]
93. Kyutoku, K.; Ioka, K.; Shibata, M. Ultrarelativistic electromagnetic counterpart to binary neutron star mergers. *Mon. Not. R. Astron. Soc.* **2014**, *437*, L6–L10. [CrossRef]
94. Rosswog, S. The dynamic ejecta of compact object mergers and eccentric collisions. *R. Soc. Lond. Philos. Trans. Ser. A* **2013**, *371*, 20272. [CrossRef] [PubMed]
95. Bauswein, A.; Goriely, S.; Janka, H.T. Systematics of Dynamical Mass Ejection, Nucleosynthesis, and Radioactively Powered Electromagnetic Signals from Neutron-star Mergers. *Astrophys. J.* **2013**, *773*, 78. [CrossRef]
96. Hotokezaka, K.; Kiuchi, K.; Kyutoku, K.; Okawa, H.; Sekiguchi, Y.I.; Shibata, M.; Taniguchi, K. Mass ejection from the merger of binary neutron stars. *Phys. Rev. D* **2013**, *87*, 024001. [CrossRef]
97. Foucart, F.; Deaton, M.B.; Duez, M.D.; O'Connor, E.; Ott, C.D.; Haas, R.; Kidder, L.E.; Pfeiffer, H.P.; Scheel, M.A.; Szilagyi, B. Neutron star-black hole mergers with a nuclear equation of state and neutrino cooling: Dependence in the binary parameters. *Phys. Rev. D* **2014**, *90*, 024026. [CrossRef]
98. Siegel, D.M.; Ciolfi, R.; Rezzolla, L. Magnetically Driven Winds from Differentially Rotating Neutron Stars and X-ray Afterglows of Short Gamma-Ray Bursts. *Astrophys. J.* **2014**, *785*, L6. [CrossRef]
99. Wanajo, S.; Sekiguchi, Y.; Nishimura, N.; Kiuchi, K.; Kyutoku, K.; Shibata, M. Production of All the r-process Nuclides in the Dynamical Ejecta of Neutron Star Mergers. *Astrophys. J.* **2014**, *789*, L39. [CrossRef]
100. Sekiguchi, Y.; Kiuchi, K.; Kyutoku, K.; Shibata, M. Dynamical mass ejection from binary neutron star mergers: Radiation-hydrodynamics study in general relativity. *Phys. Rev. D* **2015**, *91*, 064059. [CrossRef]
101. Radice, D.; Galeazzi, F.; Lippuner, J.; Roberts, L.F.; Ott, C.D.; Rezzolla, L. Dynamical Mass Ejection from Binary Neutron Star Mergers. *Mon. Not. R. Astron. Soc.* **2016**, *460*, 3255–3271. [CrossRef]
102. Sekiguchi, Y.; Kiuchi, K.; Kyutoku, K.; Shibata, M.; Taniguchi, K. Dynamical mass ejection from the merger of asymmetric binary neutron stars: Radiation-hydrodynamics study in general relativity. *Phys. Rev. D* **2016**, *93*, 124046. [CrossRef]
103. Lehner, L.; Liebling, S.L.; Palenzuela, C.; Caballero, O.L.; O'Connor, E.; Anderson, M.; Neilsen, D. Unequal mass binary neutron star mergers and multimessenger signals. *Class. Quantum Gravity* **2016**, *33*, 184002. [CrossRef]
104. Siegel, D.M.; Metzger, B.D. Three-Dimensional General-Relativistic Magnetohydrodynamic Simulations of Remnant Accretion Disks from Neutron Star Mergers: Outflows and r-Process Nucleosynthesis. *Phys. Rev. Lett.* **2017**, *119*, 231102. [CrossRef] [PubMed]
105. Dietrich, T.; Ujevic, M.; Tichy, W.; Bernuzzi, S.; Brügmann, B. Gravitational waves and mass ejecta from binary neutron star mergers: Effect of the mass ratio. *Phys. Rev. D* **2017**, *95*, 024029. [CrossRef]
106. Bovard, L.; Martin, D.; Guercilena, F.; Arcones, A.; Rezzolla, L.; Korobkin, O. On r-process nucleosynthesis from matter ejected in binary neutron star mergers. *Phys. Rev. D* **2017**, *96*, 124005. [CrossRef]
107. Fujibayashi, S.; Sekiguchi, Y.; Kiuchi, K.; Shibata, M. Properties of Neutrino-driven Ejecta from the Remnant of a Binary Neutron Star Merger: Pure Radiation Hydrodynamics Case. *Astrophys. J.* **2017**, *846*, 114. [CrossRef]
108. Fujibayashi, S.; Kiuchi, K.; Nishimura, N.; Sekiguchi, Y.; Shibata, M. Mass Ejection from the Remnant of a Binary Neutron Star Merger: Viscous-radiation Hydrodynamics Study. *Astrophys. J.* **2018**, *860*, 64. [CrossRef]
109. Kastaun, W.; Galeazzi, F. Properties of hypermassive neutron stars formed in mergers of spinning binaries. *Phys. Rev. D* **2015**, *91*, 064027. [CrossRef]
110. Kastaun, W.; Ciolfi, R.; Endrizzi, A.; Giacomazzo, B. Structure of stable binary neutron star merger remnants: Role of initial spin. *Phys. Rev. D* **2017**, *96*, 043019. [CrossRef]
111. Hanauske, M.; Takami, K.; Bovard, L.; Rezzolla, L.; Font, J.A.; Galeazzi, F.; Stöcker, H. Rotational properties of hypermassive neutron stars from binary mergers. *Phys. Rev. D* **2017**, *96*, 043004. [CrossRef]

112. Rasio, F.; Shapiro, S. TOPICAL REVIEW: Coalescing binary neutron stars. *Class. Quantum Gravity* **1999**, *16*, R1–R29. [CrossRef]
113. Rosswog, S.; Ramirez-Ruiz, E.; Davies, M.B. High-resolution calculations of merging neutron stars—III. Gamma-ray bursts. *Mon. Not. R. Astron. Soc.* **2003**, *345*, 1077–1090. [CrossRef]
114. Price, D.J.; Rosswog, S. Producing Ultrastrong Magnetic Fields in Neutron Star Mergers. *Science* **2006**, *312*, 719–722. [CrossRef] [PubMed]
115. Liu, Y.T.; Shapiro, S.L.; Etienne, Z.B.; Taniguchi, K. General relativistic simulations of magnetized binary neutron star mergers. *Phys. Rev. D* **2008**, *78*, 024012. [CrossRef]
116. Anderson, M.; Hirschmann, E.W.; Lehner, L.; Liebling, S.L.; Motl, P.M.; Neilsen, D.; Palenzuela, C.; Tohline, J.E. Magnetized Neutron-Star Mergers and Gravitational-Wave Signals. *Phys. Rev. Lett.* **2008**, *100*, 191101. [CrossRef] [PubMed]
117. Giacomazzo, B.; Rezzolla, L.; Baiotti, L. Accurate evolutions of inspiralling and magnetized neutron stars: Equal-mass binaries. *Phys. Rev. D* **2011**, *83*, 044014. [CrossRef]
118. Giacomazzo, B.; Zrake, J.; Duffell, P.C.; MacFadyen, A.I.; Perna, R. Producing Magnetar Magnetic Fields in the Merger of Binary Neutron Stars. *Astrophys. J.* **2015**, *809*, 39. [CrossRef]
119. Dionysopoulou, K.; Alic, D.; Rezzolla, L. General-relativistic resistive-magnetohydrodynamic simulations of binary neutron stars. *Phys. Rev. D* **2015**, *92*, 084064. [CrossRef]
120. Palenzuela, C.; Liebling, S.L.; Neilsen, D.; Lehner, L.; Caballero, O.L.; O'Connor, E.; Anderson, M. Effects of the microphysical equation of state in the mergers of magnetized neutron stars with neutrino cooling. *Phys. Rev. D* **2015**, *92*, 044045. [CrossRef]
121. Kiuchi, K.; Kyutoku, K.; Sekiguchi, Y.; Shibata, M.; Wada, T. High resolution numerical relativity simulations for the merger of binary magnetized neutron stars. *Phys. Rev. D* **2014**, *90*, 041502. [CrossRef]
122. Obergaulinger, M.; Aloy, M.A.; Müller, E. Local simulations of the magnetized Kelvin-Helmholtz instability in neutron-star mergers. *Astron. Astrophys.* **2010**, *515*, A30. [CrossRef]
123. Zrake, J.; MacFadyen, A.I. Magnetic Energy Production by Turbulence in Binary Neutron Star Mergers. *Astrophys. J.* **2013**, *769*, L29. [CrossRef]
124. Kiuchi, K.; Cerdá-Durán, P.; Kyutoku, K.; Sekiguchi, Y.; Shibata, M. Efficient magnetic-field amplification due to the Kelvin-Helmholtz instability in binary neutron star mergers. *Phys. Rev. D* **2015**, *92*, 124034. [CrossRef]
125. Kiuchi, K.; Kyutoku, K.; Sekiguchi, Y.; Shibata, M. Global simulations of strongly magnetized remnant massive neutron stars formed in binary neutron star mergers. *Phys. Rev. D* **2018**, *97*, 124039. [CrossRef]
126. Harutyunyan, A.; Nathanail, A.; Rezzolla, L.; Sedrakian, A. Electrical Resistivity and Hall Effect in Binary Neutron-Star Mergers. *arXiv* **2018**, arXiv:1803.09215.
127. Moesta, P.; Mundim, B.; Faber, J.; Noble, S.; Bode, T.; Haas, R.; Loeffler, F.; Ott, C.; Reisswig, C.; Schnetter, E. General relativistic magneto-hydrodynamics with the Einstein Toolkit. In Proceedings of the 2013 APS Meeting Abstracts, Baltimore, MD, USA, 18–22 March 2013; p. 10001.
128. Rembiasz, T.; Guilet, J.; Obergaulinger, M.; Cerdá-Durán, P.; Aloy, M.A.; Müller, E. On the maximum magnetic field amplification by the magnetorotational instability in core-collapse supernovae. *Mon. Not. R. Astron. Soc.* **2016**, *460*, 3316–3334. [CrossRef]
129. Zrake, J.; MacFadyen, A.I. Spectral and Intermittency Properties of Relativistic Turbulence. *Astrophys. J.* **2013**, *763*, L12. [CrossRef]
130. Rezzolla, L.; Giacomazzo, B.; Baiotti, L.; Granot, J.; Kouveliotou, C.; Aloy, M.A. The Missing Link: Merging Neutron Stars Naturally Produce Jet-like Structures and Can Power Short Gamma-ray Bursts. *Astrophys. J. Lett.* **2011**, *732*, L6. [CrossRef]
131. Blandford, R.D.; Znajek, R.L. Electromagnetic extraction of energy from Kerr black holes. *Mon. Not. R. Astron. Soc.* **1977**, *179*, 433–456. [CrossRef]
132. Contopoulos, J. A Simple Type of Magnetically Driven Jets: An Astrophysical Plasma Gun. *Astrophys. J.* **1995**, *450*, 616. [CrossRef]
133. Ruiz, M.; Lang, R.N.; Paschalidis, V.; Shapiro, S.L. Binary Neutron Star Mergers: A Jet Engine for Short Gamma-Ray Bursts. *Astrophys. J. Lett.* **2016**, *824*, L6. [CrossRef] [PubMed]
134. Kawamura, T.; Giacomazzo, B.; Kastaun, W.; Ciolfi, R.; Endrizzi, A.; Baiotti, L.; Perna, R. Binary neutron star mergers and short gamma-ray bursts: Effects of magnetic field orientation, equation of state, and mass ratio. *Phys. Rev. D* **2016**, *94*, 064012. [CrossRef]

135. Endrizzi, A.; Ciolfi, R.; Giacomazzo, B.; Kastaun, W.; Kawamura, T. General relativistic magnetohydrodynamic simulations of binary neutron star mergers with the APR4 equation of state. *Class. Quantum Gravity* **2016**, *33*, 164001. [CrossRef]
136. Ruffert, M.; Janka, H.T. Gamma-ray bursts from accreting black holes in neutron star mergers. *Astron. Astrophys.* **1999**, *344*, 573–606.
137. Just, O.; Obergaulinger, M.; Janka, H.T.; Bauswein, A.; Schwarz, N. Neutron-star Merger Ejecta as Obstacles to Neutrino-powered Jets of Gamma-Ray Bursts. *Astrophys. J. Lett.* **2016**, *816*, L30. [CrossRef]
138. Komissarov, S.S. Direct numerical simulations of the Blandford-Znajek effect. *Mon. Not. R. Astron. Soc.* **2001**, *326*, L41–L44. [CrossRef]
139. Komissarov, S.S.; Barkov, M.; Lyutikov, M. Tearing instability in relativistic magnetically dominated plasmas. *Mon. Not. R. Astron. Soc.* **2007**, *374*, 415–426. [CrossRef]
140. Nathanail, A.; Contopoulos, I. Black Hole Magnetospheres. *Astrophys. J.* **2014**, *788*, 186. [CrossRef]
141. Gralla, S.E.; Lupsasca, A.; Rodriguez, M.J. Electromagnetic jets from stars and black holes. *Phys. Rev. D* **2016**, *93*, 044038. [CrossRef]
142. Shapiro, S.L. Black holes, disks, and jets following binary mergers and stellar collapse: The narrow range of electromagnetic luminosities and accretion rates. *Phys. Rev. D* **2017**, *95*, 101303. [CrossRef] [PubMed]
143. Shibata, M.; Taniguchi, K.; Uryū, K. Merger of binary neutron stars of unequal mass in full general relativity. *Phys. Rev. D* **2003**, *68*, 084020. [CrossRef]
144. Shibata, M.; Taniguchi, K. Merger of binary neutron stars to a black hole: Disk mass, short gamma-ray bursts, and quasinormal mode ringing. *Phys. Rev. D* **2006**, *73*, 064027. [CrossRef]
145. Baiotti, L.; Giacomazzo, B.; Rezzolla, L. Accurate evolutions of inspiralling neutron-star binaries: Prompt and delayed collapse to a black hole. *Phys. Rev. D* **2008**, *78*, 084033. [CrossRef]
146. Rezzolla, L.; Macedo, R.P.; Jaramillo, J.L. Understanding the 'anti-kick' in the merger of binary black holes. *Phys. Rev. Lett.* **2010**, *104*, 221101. [CrossRef] [PubMed]
147. Margalit, B.; Metzger, B.D.; Beloborodov, A.M. Does the Collapse of a Supramassive Neutron Star Leave a Debris Disk? *Phys. Rev. Lett.* **2015**, *115*, 171101. [CrossRef] [PubMed]
148. Camelio, G.; Dietrich, T.; Rosswog, S. Disk formation in the collapse of supramassive neutron stars. *arXiv* **2018**, arXiv:1806.07775.
149. Shakura, N.I.; Sunyaev, R.A. Black holes in binary systems. Observational appearance. *Astron. Astrophys.* **1973**, *24*, 337–355.
150. Falcke, H.; Rezzolla, L. Fast radio bursts: The last sign of supramassive neutron stars. *Astron. Astrophys.* **2014**, *562*, A137. [CrossRef]
151. Shibata, M.; Uryū, K. Gravitational Waves from Merger of Binary Neutron Stars in Fully General Relativistic Simulation. *Prog. Theor. Phys.* **2002**, *107*, 265–303. [CrossRef]
152. Hotokezaka, K.; Kyutoku, K.; Okawa, H.; Shibata, M.; Kiuchi, K. Binary neutron star mergers: Dependence on the nuclear equation of state. *Phys. Rev. D* **2011**, *83*, 124008. [CrossRef]
153. Bauswein, A.; Baumgarte, T.W.; Janka, H.T. Prompt Merger Collapse and the Maximum Mass of Neutron Stars. *Phys. Rev. Lett.* **2013**, *111*, 131101. [CrossRef] [PubMed]
154. Shibata, M.; Taniguchi, K.; Uryū, K. Merger of binary neutron stars with realistic equations of state in full general relativity. *Phys. Rev. D* **2005**, *71*, 084021. [CrossRef]
155. Oechslin, R.; Janka, H.T.; Marek, A. Relativistic neutron star merger simulations with non-zero temperature equations of state. I. Variation of binary parameters and equation of state. *Astron. Astrophys.* **2007**, *467*, 395–409. [CrossRef]
156. Studzińska, A.M.; Kucaba, M.; Gondek-Rosińska, D.; Villain, L.; Ansorg, M. Effect of the equation of state on the maximum mass of differentially rotating neutron stars. *Mon. Not. R. Astron. Soc.* **2016**, *463*, 2667–2679. [CrossRef]
157. Zhang, C. M.; Wang, J.; Zhao, Y. H.; Yin, H. X.; Song, L. M.; Menezes, D. P.; Wickramasinghe, D. T.; Ferrario, L.; Chardonnet, P. Study of measured pulsar masses and their possible conclusions. *Astron. Astrophys.* **2011**, *527*, A83. [CrossRef]
158. Ruiz, M.; Shapiro, S.L. General relativistic magnetohydrodynamics simulations of prompt-collapse neutron star mergers: The absence of jets. *Phys. Rev. D* **2017**, *96*, 084063. [CrossRef] [PubMed]
159. Lorimer, D.R.; Bailes, M.; McLaughlin, M.A.; Narkevic, D.J.; Crawford, F. A Bright Millisecond Radio Burst of Extragalactic Origin. *Science* **2007**, *318*, 777. [CrossRef] [PubMed]

160. Rane, A.; Lorimer, D. Fast Radio Bursts. *J. Astrophys. Astron.* **2017**, *38*, 55. [CrossRef]
161. Most, E.R.; Nathanail, A.; Rezzolla, L. Electromagnetic emission from blitzars and its impact on non-repeating fast radio bursts. *arXiv* **2018**, arXiv:1801.05705.
162. Paschalidis, V.; Ruiz, M. Are fast radio bursts the most likely electromagnetic counterpart of neutron star mergers resulting in prompt collapse? *arXiv* **2018**, arXiv:1808.04822.
163. Rowlinson, A.; O'Brien, P.T.; Metzger, B.D.; Tanvir, N.R.; Levan, A.J. Signatures of magnetar central engines in short GRB light curves. *Mon. Not. R. Astron. Soc.* **2013**, *430*, 1061–1087. [CrossRef]
164. Zhang, B.; Mészáros, P. Gamma-Ray Burst Afterglow with Continuous Energy Injection: Signature of a Highly Magnetized Millisecond Pulsar. *Astrophys. J.* **2001**, *552*, L35–L38. [CrossRef]
165. Gao, W.H.; Fan, Y.Z. Short-living Supermassive Magnetar Model for the Early X-ray Flares Following Short GRBs. *Chin. J. Astron. Astrophys.* **2006**, *6*, 513–516. [CrossRef]
166. Fan, Y.Z.; Xu, D. The X-ray afterglow flat segment in short GRB 051221A: Energy injection from a millisecond magnetar? *Mon. Not. R. Astron. Soc.* **2006**, *372*, L19–L22. [CrossRef]
167. Metzger, B.D.; Piro, A.L.; Quataert, E. Time-dependent models of accretion discs formed from compact object mergers. *Mon. Not. R. Astron. Soc.* **2008**, *390*, 781–797. [CrossRef]
168. Metzger, B.D.; Martínez-Pinedo, G.; Darbha, S.; Quataert, E.; Arcones, A.; Kasen, D.; Thomas, R.; Nugent, P.; Panov, I.V.; Zinner, N.T. Electromagnetic counterparts of compact object mergers powered by the radioactive decay of r-process nuclei. *Mon. Not. R. Astron. Soc.* **2010**, *406*, 2650–2662. [CrossRef]
169. Giacomazzo, B.; Perna, R. Formation of Stable Magnetars from Binary Neutron Star Mergers. *Astrophys. J.* **2013**, *771*, L26. [CrossRef]
170. Dall'Osso, S.; Giacomazzo, B.; Perna, R.; Stella, L. Gravitational Waves from Massive Magnetars Formed in Binary Neutron Star Mergers. *Astrophys. J.* **2015**, *798*, 25. [CrossRef]
171. Rezzolla, L.; Kumar, P. A Novel Paradigm for Short Gamma-Ray Bursts with Extended X-ray Emission. *Astrophys. J.* **2015**, *802*, 95. [CrossRef]
172. Ciolfi, R.; Siegel, D.M. Short Gamma-Ray Bursts in the "Time-reversal" Scenario. *Astrophys. J.* **2015**, *798*, L36. [CrossRef]
173. Spitkovsky, A. Time-dependent Force-free Pulsar Magnetospheres: Axisymmetric and Oblique Rotators. *Astrophys. J. Lett.* **2006**, *648*, L51–L54. [CrossRef]
174. Contopoulos, I.; Spitkovsky, A. Revised Pulsar Spin-down. *Astrophys. J.* **2006**, *643*, 1139–1145. [CrossRef]
175. Thompson, T.A.; Burrows, A.; Pinto, P.A. Shock Breakout in Core-Collapse Supernovae and Its Neutrino Signature. *Astrophys. J.* **2003**, *592*, 434–456. [CrossRef]
176. Komissarov, S.S. Multidimensional numerical scheme for resistive relativistic magnetohydrodynamics. *Mon. Not. R. Astron. Soc.* **2007**, *382*, 995–1004. [CrossRef]
177. Thompson, T.A.; ud-Doula, A. High-entropy ejections from magnetized proto-neutron star winds: Implications for heavy element nucleosynthesis. *Mon. Not. R. Astron. Soc.* **2018**, *476*, 5502–5515. [CrossRef]
178. Uzdensky, D.A. Force-Free Magnetosphere of an Accretion Disk-Black Hole System. II. Kerr Geometry. *Astrophys. J.* **2005**, *620*, 889–904. [CrossRef]
179. Parfrey, K.; Giannios, D.; Beloborodov, A.M. Black hole jets without large-scale net magnetic flux. *Mon. Not. R. Astron. Soc.* **2015**, *446*, L61–L65. [CrossRef]
180. Contopoulos, I.; Nathanail, A.; Katsanikas, M. The Cosmic Battery in Astrophysical Accretion Disks. *Astrophys. J.* **2015**, *805*, 105. [CrossRef]
181. Nathanail, A.; Contopoulos, I. Are ultralong gamma-ray bursts powered by black holes spinning down? *Mon. Not. R. Astron. Soc.* **2015**, *453*, L1–L5. [CrossRef]
182. Nathanail, A.; Strantzalis, A.; Contopoulos, I. The rapid decay phase of the afterglow as the signature of the Blandford-Znajek mechanism. *Mon. Not. R. Astron. Soc.* **2016**, *455*, 4479–4486. [CrossRef]
183. Nathanail, A. An Explosion is Triggered by the Late Collapse of the Compact Remnant from a Neutron Star Merger. *Astrophys. J.* **2018**, *864*, 4. [CrossRef]
184. Nakar, E.; Piran, T. Implications of the radio and X-ray emission that followed GW170817. *Mon. Not. R. Astron. Soc.* **2018**, *478*, 407–415. [CrossRef]
185. Blandford, R.D.; McKee, C.F. Fluid dynamics of relativistic blast waves. *Phys. Fluids* **1976**, *19*, 1130–1138. [CrossRef]

186. Ramirez-Ruiz, E.; Celotti, A.; Rees, M.J. Events in the life of a cocoon surrounding a light, collapsar jet. *Mon. Not. R. Astron. Soc.* **2002**, *337*, 1349–1356. [CrossRef]
187. Morsony, B.J.; Lazzati, D.; Begelman, M.C. Temporal and Angular Properties of Gamma-Ray Burst Jets Emerging from Massive Stars. *Astrophys. J.* **2007**, *665*, 569–598. [CrossRef]
188. Lazzati, D.; Morsony, B.J.; Begelman, M.C. Short-duration Gamma-ray Bursts From Off-axis Collapsars. *Astrophys. J.* **2010**, *717*, 239–244. [CrossRef]
189. Mizuta, A.; Aloy, M.A. Angular Energy Distribution of Collapsar-Jets. *arXiv* **2008**, arXiv:0812.4813.
190. López-Cámara, D.; Morsony, B.J.; Begelman, M.C.; Lazzati, D. Three-dimensional Adaptive Mesh Refinement Simulations of Long-duration Gamma-Ray Burst Jets inside Massive Progenitor Stars. *Astrophys. J.* **2013**, *767*, 19. [CrossRef]
191. Mizuta, A.; Ioka, K. Opening Angles of Collapsar Jets. *Astrophys. J.* **2013**, *777*, 162. [CrossRef]
192. Nagakura, H.; Hotokezaka, K.; Sekiguchi, Y.; Shibata, M.; Ioka, K. Jet Collimation in the Ejecta of Double Neutron Star Mergers: A New Canonical Picture of Short Gamma-Ray Bursts. *Astrophys. J.* **2014**, *784*, L28. [CrossRef]
193. Murguia-Berthier, A.; Montes, G.; Ramirez-Ruiz, E.; De Colle, F.; Lee, W.H. Necessary Conditions for Short Gamma-Ray Burst Production in Binary Neutron Star Mergers. *Astrophys. J.* **2014**, *788*, L8. [CrossRef]
194. Murguia-Berthier, A.; Ramirez-Ruiz, E.; Montes, G.; De Colle, F.; Rezzolla, L.; Rosswog, S.; Takami, K.; Perego, A.; Lee, W.H. The Properties of Short Gamma-Ray Burst Jets Triggered by Neutron Star Mergers. *Astrophys. J. Lett.* **2016**, *835*, L34. [CrossRef]
195. Nagakura, H.; Ito, H.; Kiuchi, K.; Yamada, S. Jet Propagations, Breakouts, and Photospheric Emissions in Collapsing Massive Progenitors of Long-duration Gamma-ray Bursts. *Astrophys. J.* **2011**, *731*, 80. [CrossRef]
196. Qian, Y.Z.; Woosley, S.E. Nucleosynthesis in Neutrino-driven Winds. I. The Physical Conditions. *Astrophys. J.* **1996**, *471*, 331. [CrossRef]
197. Rosswog, S.; Ramirez-Ruiz, E. Jets, winds and bursts from coalescing neutron stars. *Mon. Not. R. Astron. Soc.* **2002**, *336*, L7–L11. [CrossRef]
198. Gehrels, N.; Ramirez-Ruiz, E.; Fox, D.B. Gamma-Ray Bursts in the Swift Era. *Annu. Rev. Astron. Astrophys.* **2009**, *47*, 567–617. [CrossRef]
199. Hotokezaka, K.; Kyutoku, K.; Tanaka, M.; Kiuchi, K.; Sekiguchi, Y.; Shibata, M.; Wanajo, S. Progenitor Models of the Electromagnetic Transient Associated with the Short Gamma Ray Burst 130603B. *Astrophys. J.* **2013**, *778*, L16. [CrossRef]
200. Duffell, P.C.; Quataert, E.; MacFadyen, A.I. A Narrow Short-duration GRB Jet from a Wide Central Engine. *Astrophys. J.* **2015**, *813*, 64. [CrossRef]
201. Rosswog, S.; Ramirez-Ruiz, E. On the diversity of short gamma-ray bursts. *Mon. Not. R. Astron. Soc.* **2003**, *343*, L36–L40. [CrossRef]
202. Bucciantini, N.; Metzger, B.D.; Thompson, T.A.; Quataert, E. Short gamma-ray bursts with extended emission from magnetar birth: Jet formation and collimation. *Mon. Not. R. Astron. Soc.* **2012**, *419*, 1537–1545. [CrossRef]
203. Bromberg, O.; Tchekhovskoy, A.; Gottlieb, O.; Nakar, E.; Piran, T. The γ-rays that accompanied GW170817 and the observational signature of a magnetic jet breaking out of NS merger ejecta. *Mon. Not. R. Astron. Soc.* **2018**, *475*, 2971–2977. [CrossRef]
204. Perego, A.; Rosswog, S.; Cabezón, R.M.; Korobkin, O.; Käppeli, R.; Arcones, A.; Liebendörfer, M. Neutrino-driven winds from neutron star merger remnants. *Mon. Not. R. Astron. Soc.* **2014**, *443*, 3134–3156. [CrossRef]
205. Nakar, E.; Piran, T. Detectable radio flares following gravitational waves from mergers of binary neutron stars. *Nature* **2011**, *478*, 82–84, [CrossRef] [PubMed]
206. Lazzati, D.; Deich, A.; Morsony, B.J.; Workman, J.C. Off-axis emission of short γ-ray bursts and the detectability of electromagnetic counterparts of gravitational-wave-detected binary mergers. *Mon. Not. R. Astron. Soc.* **2017**, *471*, 1652–1661. [CrossRef]
207. Nakar, E.; Piran, T. The Observable Signatures of GRB Cocoons. *Astrophys. J.* **2017**, *834*, 28. [CrossRef]

208. Bromberg, O.; Nakar, E.; Piran, T.; Sari, R. The Propagation of Relativistic Jets in External Media. *Astrophys. J.* **2011**, *740*, 100. [CrossRef]
209. Lazzati, D.; López-Cámara, D.; Cantiello, M.; Morsony, B.J.; Perna, R.; Workman, J.C. Off-axis Prompt X-ray Transients from the Cocoon of Short Gamma-Ray Bursts. *Astrophys. J. Lett.* **2017**, *848*, L6. [CrossRef]
210. Gottlieb, O.; Nakar, E.; Piran, T. The cocoon emission—An electromagnetic counterpart to gravitational waves from neutron star mergers. *Mon. Not. R. Astron. Soc.* **2018**, *473*, 576–584. [CrossRef]
211. Kathirgamaraju, A.; Barniol Duran, R.; Giannios, D. Off-axis short GRBs from structured jets as counterparts to GW events. *Mon. Not. R. Astron. Soc.* **2018**, *473*, L121–L125. [CrossRef]
212. Gottlieb, O.; Nakar, E.; Piran, T.; Hotokezaka, K. A cocoon shock breakout as the origin of the γ-ray emission in GW170817. *Mon. Not. R. Astron. Soc.* **2018**, *479*, 588–600. [CrossRef]
213. Piran, T.; Nakar, E.; Rosswog, S. The electromagnetic signals of compact binary mergers. *Mon. Not. R. Astron. Soc.* **2013**, *430*, 2121–2136. [CrossRef]
214. Hotokezaka, K.; Piran, T. Mass ejection from neutron star mergers: Different components and expected radio signals. *Mon. Not. R. Astron. Soc.* **2015**, *450*, 1430–1440. [CrossRef]
215. van Eerten, H.J.; MacFadyen, A.I. Synthetic Off-axis Light Curves for Low-energy Gamma-Ray Bursts. *Astrophys. J. Lett.* **2011**, *733*, L37. [CrossRef]
216. van Eerten, H.; van der Horst, A.; MacFadyen, A. Gamma-Ray Burst Afterglow Broadband Fitting Based Directly on Hydrodynamics Simulations. *Astrophys. J.* **2012**, *749*, 44. [CrossRef]
217. Lazzati, D.; Perna, R.; Morsony, B.J.; López-Cámara, D.; Cantiello, M.; Ciolfi, R.; giacomazzo, B.; Workman, J.C. Late time afterglow observations reveal a collimated relativistic jet in the ejecta of the binary neutron star merger GW170817. *arXiv* **2017**, arXiv:1712.03237.
218. Xie, X.; Zrake, J.; MacFadyen, A. Numerical simulations of the jet dynamics and synchrotron radiation of binary neutron star merger event GW170817/GRB170817A. *arXiv* **2018**, arXiv:1804.09345.
219. Duffell, P.C.; Quataert, E.; Kasen, D.; Klion, H. Jet Dynamics in Compact Object Mergers: GW170817 Likely had a Successful Jet. *arXiv* **2018**, arXiv:1806.10616.
220. Gill, R.; Granot, J. Afterglow Imaging and Polarization of Misaligned Structured GRB Jets and Cocoons: Breaking the Degeneracy in GRB 170817A. *Mon. Not. R. Astron. Soc.* **2018**, *478*, 4128–4141. [CrossRef]
221. Nakar, E.; Gottlieb, O.; Piran, T.; Kasliwal, M.M.; Hallinan, G. From γ to Radio—The Electromagnetic Counterpart of GW 170817. *arXiv* **2018**, arXiv:1803.07595.
222. Zrake, J.; Xie, X.; MacFadyen, A. Radio sky maps of the GRB 170817A afterglow from simulations. *arXiv* **2018**, arXiv:1806.06848.
223. Granot, J.; De Colle, F.; Ramirez-Ruiz, E. Off-axis afterglow light curves and images from 2D hydrodynamic simulations of double-sided GRB jets in a stratified external medium. *arXiv* **2018**, arXiv:1803.05856.
224. Salafia, O.S.; Ghisellini, G.; Ghirlanda, G.; Colpi, M. GRB170817A: A giant flare from a jet-less double neutron-star merger? *arXiv* **2017**, arXiv:1711.03112.
225. Salafia, O.S.; Ghisellini, G.; Ghirlanda, G. Jet-driven and jet-less fireballs from compact binary mergers. *Mon. Not. R. Astron. Soc.* **2018**, *474*, L7–L11. [CrossRef]
226. Tong, H.; Yu, C.; Huang, L. A magnetically driven origin for the low luminosity GRB 170817A associated with GW170817. *Res. Astron. Astrophys.* **2018**, *18*, 67. [CrossRef]
227. Lamb, G.P.; Kobayashi, S. GRB 170817A as a jet counterpart to gravitational wave trigger GW 170817. *arXiv* **2017**, arXiv:1710.05857.
228. Ziaeepour, H. Prompt gamma-ray emission of GRB 170817A associated with GW 170817: A consistent picture. *Mon. Not. R. Astron. Soc.* **2018**, *478*, 3233–3252, arXiv:1801.06124.
229. Veres, P.; Mészáros, P.; Goldstein, A.; Fraija, N.; Connaughton, V.; Burns, E.; Preece, R.D.; Hamburg, R.; Wilson-Hodge, C.A.; Briggs, M.S.; et al. Gamma-ray burst models in light of the GRB 170817A—GW170817 connection. *arXiv* **2018**, arXiv:1802.07328.
230. Beniamini, P.; Petropoulou, M.; Barniol Duran, R.; Giannios, D. A lesson from GW170817: Most neutron star mergers result in tightly collimated successful GRB jets. *arXiv* **2018**, arXiv:1808.04831.
231. Hotokezaka, K.; Kiuchi, K.; Shibata, M.; Nakar, E.; Piran, T. Synchrotron radiation from the fast tail of dynamical ejecta of neutron star mergers. *arXiv* **2018**, arXiv:1803.00599.
232. Yamazaki, R.; Ioka, K.; Nakamura, T. Prompt emission from the counter jet of a short gamma-ray burst. *Prog. Theor. Exp. Phys.* **2018**, *2018*, 033E01. [CrossRef]

233. Fan, X.; Messenger, C.; Heng, I.S. Probing Intrinsic Properties of Short Gamma-Ray Bursts with Gravitational Waves. *Phys. Rev. Lett.* **2017**, *119*, 181102. [CrossRef] [PubMed]
234. Metzger, B.D.; Beniamini, P.; Giannios, D. Effects of Fallback Accretion on Protomagnetar Outflows in Gamma-Ray Bursts and Superluminous Supernovae. *Astrophys. J.* **2018**, *857*, 95. [CrossRef]

© 2018 by the author. Licensee MDPI, Basel, Switzerland. This article is an open access article distributed under the terms and conditions of the Creative Commons Attribution (CC BY) license (http://creativecommons.org/licenses/by/4.0/).

Article

The Rate of Short-Duration Gamma-Ray Bursts in the Local Universe

Soheb Mandhai [1,*,†], Nial Tanvir [1,*,†], Gavin Lamb [1,†], Andrew Levan [2] and David Tsang [3]

[1] Department of Physics and Astronomy, University of Leicester, University Road, Leicester LE1 7RH, UK; gpl6@leicester.ac.uk
[2] Department of Physics, University of Warwick, Coventry CV4 7AL, UK; A.J.Levan@warwick.ac.uk
[3] Department of Physics, University of Bath, Claverton Down, Bath BA2 7AY, UK; D.Tsang@bath.ac.uk
* Correspondence: sfm13@leicester.ac.uk (S.M.); nrt3@leicester.ac.uk (N.T.)
† These authors contributed equally to this work.

Received: 31 October 2018; Accepted: 27 November 2018; Published: 30 November 2018

Abstract: Following the faint gamma-ray burst, GRB 170817A, coincident with a gravitational wave-detected binary neutron star merger at $d \sim 40\,\mathrm{Mpc}$, we consider the constraints on a local population of faint short duration GRBs (defined here broadly as $T_{90} < 4\,\mathrm{s}$). We review proposed low-redshift short-GRBs and consider statistical limits on a $d \lesssim 200\,\mathrm{Mpc}$ population using Swift/Burst Alert Telescope (BAT), Fermi/Gamma-ray Burst Monitor (GBM), and Compton Gamma-Ray Observatory (CGRO) Burst and Transient Source Experiment (BATSE) GRBs. Swift/BAT short-GRBs give an upper limit for the all-sky rate of $<4\,\mathrm{y}^{-1}$ at $d < 200\,\mathrm{Mpc}$, corresponding to $<5\%$ of SGRBs. Cross-correlation of selected CGRO/BATSE and Fermi/GBM GRBs with $d < 100\,\mathrm{Mpc}$ galaxy positions returns a weaker constraint of $\sim 12\,\mathrm{y}^{-1}$. A separate search for correlations due to SGR giant flares in nearby ($d < 11\,\mathrm{Mpc}$) galaxies finds an upper limit of $<3\,\mathrm{y}^{-1}$. Our analysis suggests that GRB 170817A-like events are likely to be rare in existing SGRB catalogues. The best candidate for an analogue remains GRB 050906, where the Swift/BAT location was consistent with the galaxy IC 0327 at $d \approx 132\,\mathrm{Mpc}$. If binary neutron star merger rates are at the high end of current estimates, then our results imply that at most a few percent will be accompanied by detectable gamma-ray flashes in the forthcoming LIGO/Virgo science runs.

Keywords: short gamma-ray bursts; physics; progenitors; host galaxies

1. Introduction

Gamma-ray bursts are classified as either long or short duration. The distinction is most clearly indicated by the time taken to receive 90% of the total gamma-ray fluence, the T_{90} of the burst. The canonical definition places bursts having $T_{90} > 2\,\mathrm{s}$ in the long-duration class and those with $T_{90} < 2\,\mathrm{s}$ in the short-duration (SGRB) class [1], although in practice, there is an overlap in the population properties, and measured duration is also detector dependent (e.g., [2]).

The short class is thought to arise predominantly during the gravitational wave-driven merger of compact binary objects, particularly binary neutron stars (NSNS) or binary stellar-mass black hole and neutron star systems (NSBH) [3–9]. The rapid accretion of material disrupted during the merger will launch an ultra-relativistic jet, which gives rise to an SGRB for observers aligned within the opening angle of the jet. The compact binary merger scenario is supported by the broad population of SGRB host galaxies, both with and without recent star formation [10], and the wide range of offsets between burst location and host [11,12]. These host offsets suggest that there is a delay between formation and merger and that such systems can sometimes receive large natal kicks from the supernovae that form the compact objects [13]. In a handful of cases, possible transients powered by nucleosynthesis in the neutron-rich ejecta—so called "kilonovae" (often called "macronovae")—have also been detected (e.g., [14–17]).

The detection of the short-duration GRB 170817A accompanying the gravitational wave-detected merger GW170817 of a binary neutron star system consolidated the idea that compact binary mergers are the progenitors of SGRBs (e.g., [18]). However, GRB 170817A was unusual when compared to previous short-bursts for which the redshift has been determined. In particular, at $d \approx 40$ Mpc, GRB 170817A was much closer and intrinsically less luminous than any previous SGRB with a securely-measured distance. Nonetheless, this discovery has revived interest in the rate of SGRBs (and events that are phenomenologically similar) in the local universe ($d \lesssim 200$ Mpc), which has particular bearing on the expected fraction of gravitational wave detections that will be accompanied by detectable gamma-ray flashes (e.g., [19]).

Previous work, looking at the spatial correlation between galaxies in the local universe and the positions in the sky of SGRBs from the Compton Gamma-Ray Observatory (CGRO) Burst and Transient Source Experiment (BATSE) catalogue concluded that as many as 10–25% of $T_{90} < 2$ s bursts could be of a local origin [20,21]. However, this correlation result was of comparatively weak statistical significance, while the poor positional information (a few degrees of precision) provided by BATSE was not sufficient to identify any given burst with a particular host galaxy. A rate of binary mergers as high as 25% would have significant implications for the expected detection rates in future GW observations and also for their contribution to the heavy element nucleosynthesis budget. As such, it is important to re-assess the local SGRB rate in light of more recent observational evidence. Since 2005, the Swift satellite has localised ~100 SGRBs with at least a few arcmin astrometric precision, with none being clearly associated with a local galaxy. This certainly suggests that the rate of local SGRBs must be below that estimated from the original cross-correlation analysis. In this paper, we revisit this question, first discussing a range of potential low-luminosity SGRB-like events, then considering the observational constraints on such populations based on the samples of bursts seen by Swift, CGRO/BATSE, Fermi/GBM, and the Inter-Planetary Network.

2. Potential Low-Redshift Short Gamma-Ray Transients

2.1. Short Gamma-Ray Bursts

The observed population of SGRBs with measured redshift spans a range from $z \sim 0.1$ (e.g., [22]) to $z > 1$ (e.g., [23,24]), based primarily on events that have been well-localised by the Neil Gehrels Swift Observatory in nearly 14 years of observation to date. Thus, it seems that the rate of on-axis SGRBs with detectable gamma-ray emission in the local universe ($z \lesssim 0.05$) is likely to be low. However, the faint-end of the cosmological SGRB luminosity function is not well constrained, and it may be that faint SGRBs without a redshift measurement contribute to a local population if they are viewed off-axis slightly outside of the jet core [25–27].

2.2. GW170817-Like Events

It is now clear, following GRB 170817A, that some NSNS mergers can also produce weak events. Since there is good evidence that our sight-line was well off the primary jet axis for GW170817 [28,29], it is more likely that the observed burst of gamma-rays in this event was produced by a shock-breakout of a cocoon, rather than being emission from internal shocks in the jet (e.g., [30–33]).

GRB 170817A has re-ignited interest in the local rate of SGRBs < 200 Mpc, approximately the sensitivity limit to NSNS mergers for the current generation of gravitational wave detectors. GRB 170817A-like transients could produce a population of faint SGRBs with durations of a few seconds. Bursts with a comparable energy to GRB 170817A could be detectable in gamma-rays at distances of ~100 Mpc with current technology, which is well matched to that of the current generation of gravitational wave interferometers.

2.3. NSNS Merger Precursors

A small fraction $\lesssim$10% of SGRBs seem to be preceded by a detectable precursor (a short and faint flash of gamma-rays) [34]. These precursors occur 1–10 s before the SGRB at a time when the pre-merger NSs are strongly interacting. Tidal deformation of the merging NS crusts that exceeds the breaking strain will create cracks that can result in the isotropic emission of gamma-rays at a comparable energy to the observed precursors [35]. Alternatively, this tidal mechanism may shatter the crust due to the excitation of a resonant mode during the periodic deformation, a resonant shattering flare (RSF) [36].

Alternative explanations for burst precursors include the breakout of a shock-wave produced by the NSNS collision [37] or a pair fireball created by magnetospheric interaction between the merging NSs [38]. A burst of gamma-rays produced by these precursor mechanisms would be emitted isotropically and will result in a faint and local population of SGRB-like transients. A population of faint and short $\leq$0.5-s gamma-ray transients could be apparent with a similar host-galaxy type association and offset as the general SGRB population.

2.4. Giant Flares from Soft Gamma-Ray Repeaters

An entirely different class of event is expected to contribute to a local population of faint SGRB-like transients, namely giant flares (GF) from soft-gamma-ray repeaters (SGRs). For the purposes of this work, we are interested in the rate of SGR GFs in external galaxies as a potential contaminant of the SGRB catalogues that we are considering.

Highly magnetised neutron stars, or magnetars, were confirmed as the origin for SGRs with the observation of several outbursts from SGR 1900+14 after a prolonged period of quiescence [39]. On extremely rare occasions, an SGR will emit a giant flare with energy $\sim$1000-times greater than a regular SGR outburst.

On 27 December 2004, such a giant flare (GF) erupted from the magnetar SGR 1806-20 [40,41]. Somewhat less powerful giant flares had previously been observed from both SGR 0526-66 and SGR 1900+14 [42]. The isotropic equivalent energy of the SGR 1806-20 flare was initially reported as 2×10^{46} erg, based on an estimated distance of 15 kpc, meaning that similar events could potentially be observable by BATSE to a distance of $\sim$30–40 Mpc [40,41]. This raised speculation that a proportion of detected SGRBs might in fact be SGR GFs in low redshift galaxies. The distance estimate for SGR 1806-20 was later revised by Bibby et al. [43] to 8.7 kpc, reducing the peak luminosity estimate by a factor of $\sim$3 and the maximum distance that such a flare would be observable to $\sim$20–25 Mpc.

With only three detections of GFs in the MW and LMC over a period of 40–50 years, SGR GFs must be reasonably rare events. Of the known SGR GFs in the Milky Way (MW) and Large Magellanic Cloud (LMC), the durations are in the range 0.2–0.5 s [41,44,45]. Although this time scale is not directly comparable to the BATSE T_{90} durations, it does suggest that SGR GFs should typically be <1 s in duration.

3. SGRBs Observed by Swift

The Neil Gehrels Swift Observatory (Swift) is dedicated to detecting and following up on gamma-ray burst events. The on-board instruments consist of the Burst Alert Telescope (BAT), the X-ray Telescope (XRT), and the UV/Optical Telescope (UVOT) [46]. The BAT instrument observes $\sim$100 GRBs per year, providing a positional accuracy of a few arcminutes within a (fully-coded) field-of-view of $\sim$1.4 sr. A burst trigger is usually followed by an automated re-pointing to bring the event location to within the fields of the narrow-field instruments, with typical slew times of 20–70 s [47]. The XRT and UVOT are then able to make high resolution follow-up observations of the GRB afterglow and reduce the positional uncertainty to an accuracy within 0.5–5 arcseconds [48,49]. These capabilities mean that Swift generally can localise GRBs much more precisely than other missions. To underline this point, if the host galaxy NGC 4993 for GRB 170817A had been in the Swift Burst Alert Telescope

(BAT) field of view when GW170817 occurred, then it would have been immediately identified as the likely host based on the low probability (≈0.03%) of finding such a bright galaxy by chance within a given BAT gamma-ray localisation circle of a few arcmin radius. A scale comparison of a Fermi GBM error region for GRB 170817A relative to a typical Swift localisation is shown in Figure 1.

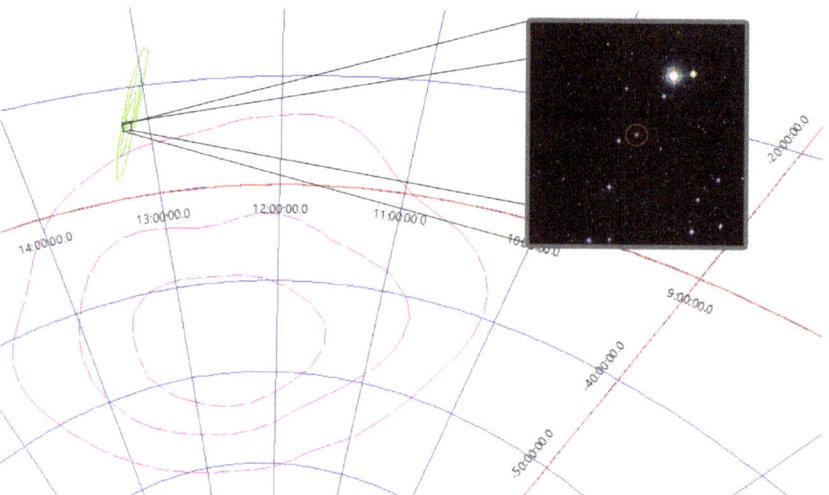

Figure 1. The GRB 170817A localisation by Fermi (magenta contours) in comparison to a typical Swift-BAT instrument localisation (red circle in zoomed panel) if the burst had been within the BAT field-of-view. The BAT region corresponds to a localisation with a radius of 3', and the inset panel is based on VLT/VIMOS imaging reported in Tanvir et al. [50]. The LIGO/Virgo localisation for GW170817 is shown as green contours, with a one-square degree box centred on the origin of the event, the galaxy NGC 4993.

To date, there have been no Swift SGRBs that have been unambiguously associated with $d < 200$ Mpc host galaxies. However, possible associations have been pointed out for GRBs 050906 [51], 070809 [52], 090417A [53], and 111020A [12] at distances $d \lesssim 400$ Mpc (see Table 1).

In order to conduct a more systematic survey, we have searched in the 2MASSRedshift Survey (2MRS) catalogue for further potential low redshift hosts. We used 2MRS as it provides a uniform coverage of galaxies over ≈91% of the sky and yields a 97.6% redshift completeness to a limiting K-band magnitude of $K = 11.75$ [54]. For the burst sample, given that an off-axis GRB is likely to have a longer duration than a canonical SGRB and that GRB 170817A was not produced by an on-axis jet and lasted ∼2 s, we consider bursts with emission duration $T_{90} < 4$ s, resulting in a sample of 157 events. In addition to this search, we also performed a visual examination of DSS-II(red) images of the regions around bursts, since some galaxies within this volume may be fainter than the 2MRS catalogue limit. We found no compelling examples beyond those already in the literature or found by the automated search.

For each burst, we found the 2MRS galaxy with the lowest impact parameter (projected distance on the plane of the sky) and within a distance of 5–200 Mpc, using galaxy distances from HyperLEDA [55] or the NASA/IPAC Extragalactic Database (NED) [56][1]. We set an upper limit for the impact parameter of 200 kpc, which allows for the possibility of binaries that received moderate natal kick velocities of

---

[1] We place a lower limit on the distance of 5 Mpc because the angular scales associated with galaxies closer than this suggest that a significant fraction of the sky is within 200 kpc (in projection) of a galaxy within this distance horizon.

~100 km s^{-1} and large binary merger times $\gtrsim 10^9$ years. Note that the impact parameter calculation is less certain for cases where there was only a BAT localisation. Finally, we removed any matches for which a more distant origin was already established from a close proximity to a higher redshift host.

Table 1. List of Swift catalogued SGRB detections that have been paired with the closest 2MASSRedshift Survey (2MRS) galaxy. The galactic distances used have been obtained from the associated references. Further bursts for which tentative host galaxies have been suggested in the literature are listed below the table break.

GRB	T_{90} (s)	Angular Separation (arcmin)	Closest Galaxy	Galaxy Type	Optical Bands (B/R) (mag)	J-Band (mag)	d (Mpc)	Impact Parameter (kpc)	E_{iso} (10^{46} ergs)
050906	0.26	2.0	IC 0328	Sc	14.0 (B)	12.2	132 [55]	77 ± 109	1.9
100213A	2.40	5.4	PGC 3087784	S0-a	14.7 (B)	11.3	78 [55]	123	39.9
111210A	2.52	6.0	NGC 4671	E	13.4 (B)	10.1	43 [55]	76	7.5
120403A	1.25	4.9	PGC 010703	Sc	14.4 (B)	12.1	133 [55]	192 ± 90	38.2
130515A	0.29	8.5	PGC 420380	S0-a	16.0 (B)	12.3	73 [57]	180	28.4
160801A	2.85	6.7	PGC 050620	Sa	15.2 (B)	12.4	59 [55]	115	10.7
070809	1.30	2.0	PGC 3082279 [52]	Sa	16.3 (B)	13.5	180 [56]	105	64.4
090417A	0.07	4.4	PGC 1022875 [53]	S0-a	15.9 (B)	13.4	360 [56]	461 ± 292	24.5
111020A	0.40	2.3	FAIRALL 1160	Sa	~14 (R)	11.7	81 [12]	54	9.4

This procedure finds potential nearby ($d < 200$ Mpc) hosts for GRBs 050906, 100213A, 111210A, 120403A, 130515A, and 160801A (full details are reported in Table 1 and shown in Figure 2). Of these, only GRB 050906 had previously been noted in the literature[2]. It is interesting to note that if this host association is correct, then its isotropic energy was comparable to that of GRB 170817A. This case is also the only one for which the host candidate is within the positional uncertainty of the burst, implying that all the other cases would require the progenitor binaries to be kicked well outside their parent galaxies if the associations were real. A high proportion of mergers occurring at large galactocentric radii would seem unlikely, for example being inconsistent with the comparatively low rate (~25%) of hostless bursts found for SGRBs in the sample studied by Fong et al. [58]. Furthermore, repeating the experiment for large numbers of random positions showed that similar apparent associations are expected to occur by chance for ~12 bursts in a sample of the size of our Swift sample. This suggests that the majority, quite possibly all, of these candidate associations, both from our 2MRS analysis and those reported previously, are likely to be chance alignments. In this regard, we note that several of the well-localised bursts from Table 1 have plausible alternative higher redshift hosts suggested in the literature, albeit that kicks would still be required to place the mergers outside the bodies of these galaxies (e.g., GRBs 070809 and 111020A in [12], GRB 111210A in [59], and GRB 130515A in [60]). Furthermore, in several cases, deep follow-up imaging places strong constraints on the luminosities of any associated kilonovae if the low redshift association were correct (e.g., GRB 050906 in [51], GRB 070809 in [52], GRB 111020A in [61], GRB 130515A in [62]).

Over the current elapsed mission duration of Swift, it has made roughly two years of all-sky observations with BAT (based on nearly 14 years in orbit and an instantaneous field of view of 10–15% of the sky). Thus, from the evidence from Swift, we conclude a limit to the all-sky rate of detectable SGRBs of <4 y^{-1} for $d < 200$ Mpc, with the likely rate being significantly less. This could only be underestimated if low redshift mergers are typically happening at very large galactocentric distances, requiring large average kicks and long merger times, or if there is a preference for very faint hosts.

---

[2] The candidate host for GRB 111210A, Fairall 1160, is brighter than the 2MRS magnitude limit, but was erroneously classified as a star in the 2MASS database, and hence not included in 2MRS.

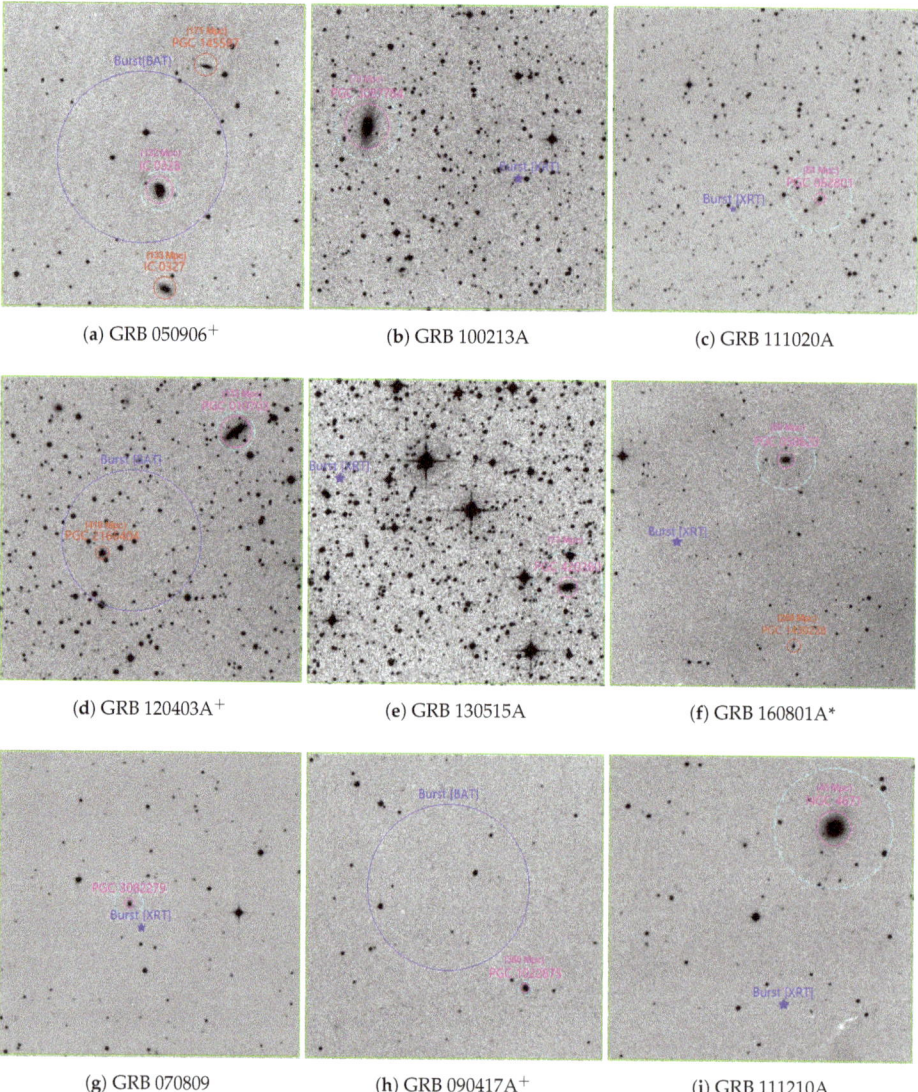

Figure 2. Positions of Swift-detected GRBs with $T_{90} < 4$ s (blue) (from Table 1) plotted on DSS2 images of the fields, with the best candidate low redshift galactic host circled in magenta. Other potential candidates at greater distances are circled in red. A projected kick distance radius of 25 kpc is highlighted by the dashed cyan circle around the favoured host candidate. The standard panel size used for each source corresponds to an on-sky area of $10' \times 10'$. * Corresponds to a $15' \times 15'$ area.
+ For events where an XRT localisation could not be obtained, either due to a delayed/absent slew (GRBs 090417A, 120403A) or a lack of an X-ray afterglow (GRB 050906), the Burst Alert Telescope (BAT) localisation is used.

4. SGRBs Observed by CGRO/BATSE and Fermi/GBM

Although Swift provides excellent positional accuracy, there have only been ~150 $T_{90} < 4\,\text{s}$ SGRBs detected over 14 years. CGRO/BATSE and Fermi/GBM have observed much larger samples of SGRBs, but with poorer localization; however, we can perform statistical analysis on the larger sample to constrain the fraction of the population that could arise from nearby galaxies.

The BATSE instrument on the space-based gamma-ray CGRO satellite continuously observed the unocculted sky from low Earth orbit, giving a field of view of $\sim 2\pi$ sr. During its nine-year lifetime, BATSE detected ~500 SGRBs with $T_{90} < 2\,\text{s}$. However, the large location uncertainties (1σ errors, typically several degrees) effectively prevented identification of the galactic hosts of these bursts. By correlating a sample of 400 BATSE SGRBs for which location errors were less than 10 degrees, with a sample of local galaxies, Tanvir et al. [20] were able to place a limit for the rate of short-duration gamma-ray bursts within $d \sim 110\,\text{Mpc}$ of ~25% of BATSE bursts. Intriguingly, this analysis showed a positive cross-correlation at a $\sim 3\sigma$ level.

The Gamma-ray Burst Monitor (GBM) instrument on the Fermi Gamma-ray Space Telescope (Fermi) has provided similar capabilities to BATSE for the past 10 years. Again, although it has observed a large sample of bursts, positional accuracy is also much lower than Swift, at a few degrees [63].

In this study, we revisit the cross-correlation analysis reported in Tanvir et al. [20] with an updated and more complete galaxy redshift catalogue (the 2MASS Redshift Survey (2MRS); [54]) and a larger sample of 782 bursts from combining both the CGRO/BATSE and Fermi/GBM catalogues. We require burst localisation error radii <10° and a $T_{90} \leq 4\,\text{s}$, and removed Fermi/GBM bursts that have been well-localised by other satellites. The localisation errors also include estimates of systematic uncertainties following Model 2 for BATSE of Briggs et al. [64] and Connaughton et al. [63] for GBM, each of which includes a core and tail systematic component. The errors were assumed to follow a Fisher distribution [64]. Distances to galaxies were taken from the HyperLEDA [55] and Cosmicflows [57] compilations.

The correlation statistic, Φ, matches each short-duration gamma-ray burst against every 2MRS galaxy within a given distance cut, summing up the likelihoods that a given burst would be found at the observed distance from the given galaxy if they were truly associated. Bursts were weighted inversely with their positional location error radii. The galaxies were also weighted according to their absolute B-band luminosity, to provide some account for the likely higher rate of binary mergers in large galaxies and galaxies with ongoing or recent star formation (cf. [58]). The correlation statistic for the real bursts was then compared to a distribution of Φ values obtained for an artificial sample containing both randomly-distributed bursts, the average value of which we denote Φ_0, as well as a fraction that were correlated with 2MRS galaxies. This approach allows us to set limits on the fraction of correlated bursts in the real sample as a function of galaxy distance cut, shown in Figure 3.

We conclude a 2σ upper limit on the fraction of all BATSE and GBM short bursts within ~100 Mpc of $\stackrel{<}{\sim}17\%$; this places an upper limit on the annual rate to be $\stackrel{<}{\sim}12\,\text{y}^{-1}$.

As can be seen from Figure 3, there is no statistically-significant evidence for non-zero correlation at any distance. Although this conclusion differs from that of Tanvir et al. [20], the two results are consistent within their 1σ error ranges, and the difference is primarily a consequence of updated samples of bursts and galaxies.

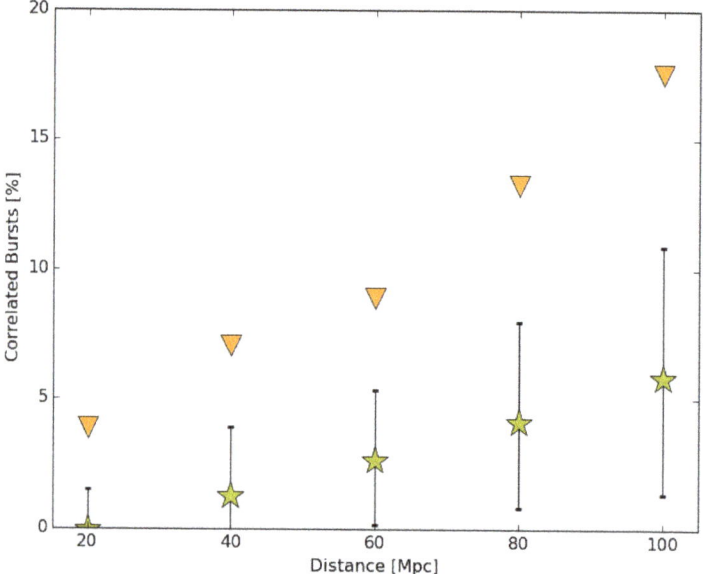

Figure 3. The correlated short-duration gamma-ray burst percentage (for $T_{90} < 4\,\text{s}$) based on the Φ/Φ_0 spread determined for each d_{gal} volume. The error bars correspond to 1σ deviation from the observed correlation ratio, Φ_{obs}/Φ_0. The yellow stars represent the non-zero correlation percentage. Orange triangles are the 2σ upper limits.

A Search for Extragalactic SGR Giant Flares

Distinguishing between transients related to flaring magnetars and those associated with merging NSs is difficult. However, SGRs are relatively young and can be expected to trace the star formation, whereas merging compact binary systems may undergo a significant delay time or be kicked substantially away from their formation site [13]. A nearby, $\lesssim$10-Mpc population of short gamma-ray transients that are associated with star-forming galaxies could indicate a population of extragalactic SGR GFs.

Within the BATSE and GBM sample of SGRBs a fraction of these bursts could be the result of SGR GFs. Using Karachentsev et al.'s [65] galaxy catalogue, which includes the star formation rate (SFR) for each galaxy, we computed the correlation for BATSE and GBM SGRBs with $T_{90} \leq 1\,\text{s}$ with galaxies <11 Mpc weighted by their SFR. The sky position for each of these bursts (orange squares for BATSE and magenta triangles for GBM) and the galaxies (blue disks) are shown in the right panel of Figure 4; the area of the galaxy marker is weighted by the SFR, so a larger area represents a higher SFR. For 477 selected BATSE and Fermi GRBs in our sample, we find a 2σ upper limit of $\sim$8%, corresponding to $\lesssim 3\,\text{y}^{-1}$, for the fraction of bursts that could be correlated with these nearby, high SFR galaxies (see the left panel of Figure 4). Once again, we note that the absence of $d < 11$-Mpc galaxies in the error regions of any Swift-detected bursts, confirming that this limit is a hard one.

A number of previous studies constrained the fraction of SGR GFs in SGRB catalogues. A sample of 76 well-detected BATSE SGRBs was analysed by Lazzati et al. [66], who concluded that only three were consistent with having black body spectra, and so candidates for extra-galactic GFs, albeit the durations of these candidates were $>1\,\text{s}$ and longer than the GFs observed in the Milky Way system.

Popov and Stern [67] argued that the rate of GFs observed in the Milky Way system would lead to the expectation that 15–25 SGR GFs from four galaxies (M82, NGC 253, NGC 4945, and M83) should have been detected during the life of the BATSE instrument. They noted that some BATSE SGRBs had positions consistent with these galaxies, but when looked at in detail, concluded that their spectra and light curves were not as expected for GFs. Furthermore, over the current elapsed mission time, Swift has made no such detections arising from M82 or similar candidates such as NGC 253, NGC 4945, and M83, despite having a somewhat reduced trigger threshold for many of these nearby galaxies.

An upper limit on the fraction of SGR GFs in the ($T_{90} < 2$ s) SGRB population at <15% was made by Nakar et al. [68] using a sample of six well-localised SGRBs. Similarly, using 47 Inter-Planetary Network (IPN) localised SGRBs, Ofek [69] checked for coincidence between bright and star-forming galaxies within 20 Mpc and the SGRB IPN sky position annulus for each burst. A single match between GRB 000420B and M74 was found, although this is likely to be a chance coincidence. By assuming an upper limit cut-off for the SGR GF isotropic energy distribution of $<10^{47}$ erg, an upper limit of <16% was found for the fraction of SGR GFs in the SGRB population. Ofek [69] placed a lower limit on the fraction at 1% based on the Galactic SGR GF rate. This range for the fraction of 1–15%, based on statistical analysis, was reported by Hurley [42].

The upper limit for the fraction of SGR GFs in the SGRB population was estimated to be <7% by Tikhomirova et al. [70] from analysis of SGRBs with relatively small localisation regions based on BATSE and IPN detections. More recently, Svinkin et al. [71] found a similar limit of <8%; they considered the evidence for extragalactic SGR GFs amongst 16 years of well-localised Konus-WIND SGRBs and found no compelling cases apart from the GRBs 051103 and 070201, discussed in Section 5.

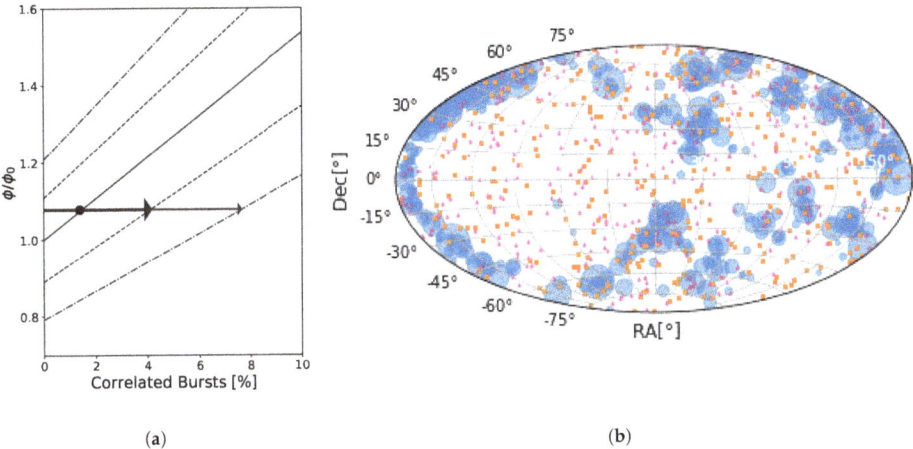

Figure 4. (a) Ratio of the weighted correlation parameter Φ to Φ_0, the value obtained for a random distribution of bursts, as a function of the fraction of simulated bursts seeded by galaxy positions (solid diagonal line). The galaxies were obtained from the compilation of Karachentsev et al. [65], which is restricted to $d_{gal} = 11$ Mpc. The dashed lines in each case correspond to the 1σ and 2σ Φ/Φ_0 ranges determined by the spread in the simulation results. The horizontal arrows are at the level of Φ/Φ_0 obtained for the real catalogue of BATSE and GBM $T_{90} < 1$ s bursts and indicate the plausible fraction that are correlated with nearby galaxies (b) The observed galaxies (blue) within 11 Mpc with star formation rates from Karachentsev et al. [65], shown alongside Burst and Transient Source Experiment (BATSE) (orange squares) and GBM (magenta triangles) SGRB detections. The area of each galactic point is scaled to reflect high star-forming environments.

5. IPN-Observed SGRBs

The Inter-Planetary Network (IPN) is the name given to a collaborative group of GRB detectors on various satellites and probes. Some of these are dedicated GRB missions, but others piggy-back on other missions. It has been operational through most of the past four decades and has nodes both at the Earth and elsewhere in the Solar System. The location of a burst can be accurately constrained by measuring the arrival time across several IPN satellites; the precision of which depends on the baseline separation [44,71,72]. The long operational life of the network, and all-sky coverage, has made it particularly successful at finding rare and bright GRB events [73]. In fact, the IPN has detected all three SGR giant flares originating in the MW/LMC [71].

IPN detections have helped provide low-z host candidates for several bursts, notably M31 for GRB 070201 [74], M81 for GRB 051103 [75], and NGC 3313 for GRB 150906B [76]. The former two were particularly bright events, adding to the case of their being at short distances. If these host associations are correct, then the isotropic equivalent energy for GRB 051103 and GRB 070201 would be 7.5×10^{46} erg and 1.5×10^{45} erg, respectively [75]. These are interestingly close to the fluence of GRB 170817A; however, the lack of coincident gravitational wave detections makes it unlikely that these bursts originated from compact binary mergers [77–79]. Based on the proximity of the potential host galaxies, it is more likely that these detections correspond to giant flares from soft gamma-ray repeaters (SGRs), particularly for GRBs 051103 and 070201.

6. Discussion and Conclusions

We have estimated the number of Swift gamma-ray bursts with $T_{90} < 4$ s that could be associated with galaxies within the Advanced-LIGO horizon. A duration cut at 4 s was adopted considering that events similar to GRB 170817A may be typically rather longer than traditional short-duration GRBs. We searched for any with minimum separation distance (or impact parameter b) of the burst from a galaxy of $b < 200$ kpc to allow for the possibility of NSNS/NSBH binaries being kicked to large distances from their hosts in some cases. In roughly two all-sky years of Swift observations, we find eight host candidates within 200 Mpc. The majority of these are likely to be chance associations, which suggests a maximum detectable SGRB all-sky rate of <4 y^{-1} at this distance. Only if typical kick distances are very large ($\gg$200 kpc) would it become difficult to identify the potential hosts for these SGRBs due to the large separation, which would potentially allow a higher rate.

Based on the impact parameter analysis of Swift bursts, the corresponding volumetric rate density is $R < 120$ Gpc^{-3} y^{-1}. This falls below the 1σ range estimated for NSNS mergers by Abbott et al. [80] of $R = 1540^{+3200}_{-1220}$ Gpc^{-3} y^{-1} based on the the single detection of GW170817 during the O1 and O2 LIGO science runs. Thus, if the true merger rate is as high as this, then only a small fraction of future GW detections is likely to be accompanied by detectable gamma-ray flashes. That could be understood as a consequence of anisotropic gamma-ray emission, since our line of sight was within 30° of the primary jet axis in the case of GW170817. On the other hand, estimates of the NSNS merger rate based on the small sample of known Milky Way double neutron stars continue to point to lower figures, albeit also with large uncertainties (e.g., $R \sim 50$ Gpc^{-3} y^{-1} in [81]).

The strongest candidate for a low-z short GRB remains GRB 050906 [51], whose BAT error box contains the galaxy IC 0328 at 132 Mpc. At this distance, its isotropic equivalent energy would be $E_{iso} = 1.9 \times 10^{46}$ erg (15–150 keV) (the bolometric correction likely yields a total energy a factor of a few higher than this), which lies in the same regime as GW170817/GRB170817A. The non-detection of a clear X-ray afterglow at the time of GRB 050906 was regarded as a likely indication that it was not a local event. However, the experience with GW170817/GRB170817A shows that off-axis afterglow emission can rise at later times at all wavelengths [82–84] at levels undetectable to the Swift-XRT [85]. Hence, GRB 050906 could well be associated with a binary merger within IC 0328, which was viewed away from its axis, although deep optical and near-IR imaging places strong limits on any emission from any accompanying kilonova.

For bursts with a poorer localisation, from CGRO/BATSE and Fermi/GBM, we follow the statistical method of Tanvir et al. [20] to measure the correlation of burst positions with those of nearby galaxies. We have found upper limits on the fraction of gamma-ray bursts with $T_{90} < 4$ s detected by BATSE and GBM that could be associated with galaxies within 100 Mpc. This analysis results in a weaker constraint, setting a 2σ upper limit for the fraction of bursts at $\lesssim 17\%$ (see Figure 3).

An extragalactic population of SGR GFs, similar to those observed in the Milky Way system, within 10–25 Mpc would appear as low-luminosity SGRBs. We have tailored a study specifically to constrain the fraction of SGR GF events masquerading as SGRBs, using galaxies within 11 Mpc, weighting them by their star formation rate [65], correlating them with bursts with a $T_{90} < 1$ s (as shown in Figure 4). This analysis places a 2σ upper limit on the fraction of such SGRBs that are due to SGR GFs within 11 Mpc of $<3\,\mathrm{y}^{-1}$. This limit is supported by the lack of Swift identified SGR GF events in nearby galaxies. In turn, given the small number statistics, it is reasonably consistent with the observation of three GFs in the Milky Way system in ~ 40 y, taking into account that the overall star formation rate in this volume is about $50\times$ that of the Milky Way system (e.g., [86]).

Another potential source of low-luminosity GRB-like events results from resonant shattering flares of neutron stars shortly before their merger in a binary system. These gamma-rays are emitted isotropically from the source and will usually not be accompanied by an SGRB beamed towards the observer. In rarer cases where the SGRB is favourably aligned, these bursts of gamma-rays could appear as a precursor to the SGRB. These short duration bursts of gamma-rays would appear similar to a population of SGR GFs, but would follow the same host distribution and offsets as the SGRB population. As it is unlikely that an SGR would have any offset from its host galaxy, due to their short lifetime and association with star-forming regions, the candidates with $T_{90} \lesssim 0.5$ s listed in Table 1 could include a population of RSFs or other precursor type transients.

Finally, we reiterate our main conclusion for the prospects of coincident gamma-ray signals with GW merger events found in the upcoming runs of the advanced generation gravitational wave detectors. The predicted NSNS detection rate during these runs remains very uncertain, but estimates range up to 50–80 per year [87]. If the true rate of mergers is as high as this, then our results suggest that only a small percentage, $<10\%$, is likely to exhibit prompt gamma-ray flashes.

Author Contributions: S.F.M., N.R.T., and G.P.L wrote the paper. N.R.T. conceived of the original analysis. S.F.M., N.R.T., and G.P.L. contributed equally to the revised analysis noted in this article. A.J.L. and D.T. contributed comments and analysis to assist in the writing of this manuscript. Breakdown: Conceptualization, S.M., N.T., and G.L.; formal analysis, S.M., N.T., and G.L.; investigation, S.M., N.T., and G.L.; methodology, S.M., N.T., and G.L.; supervision, N.T.; validation, N.T., G.L., A.L., and D.T.; writing, original draft, S.M., N.T., G.L., A.L., and D.T.

Funding: This research received no external funding.

Acknowledgments: The authors thank Andrew Blain for useful discussions. We would like to extend our gratitude to the reviewers of this paper for their useful feedback and comments. We acknowledge the usage of the following databases: HyperLEDA; Extragalactic Distance Database; NASA/IPAC Extragalactic Database; Two Mass Redshift Survey. For the images used in Figure 2, we acknowledge the usage of the Digitized Sky Survey produced at the Space Telescope Science Institute under U.S. Government Grant NAG W-2166. The images of these surveys are based on photographic data obtained using the Oschin Schmidt Telescope on Palomar Mountain and the U.K. Schmidt Telescope. The plates were processed into the present compressed digital form with the permission of these institutions. S.F.M. is supported by a PhD studentship funded by the College of Science and Engineering at the University of Leicester; G.P.L. is supported by STFC grants; N.R.T and A.J.L. acknowledge support through ERC Grant 725246 TEDE.

Conflicts of Interest: The authors declare no conflict of interest.

Abbreviations

The following abbreviations are used in this manuscript:

SGRB   Short-Duration Gamma-ray Burst
NSNS   Binary Neutron Stars
NS   Neutron Star
BHNS   Black Hole-Neutron Star Pair

SGR   Soft Gamma-ray Tepeater
GF   Giant Flare
RSF   Resonant Shattering Flare
SFR   Star Formation Rate
MW   Milky Way
LMC   Large Magellanic Cloud
DSS   Digitized Sky Survey
2MASS   Two Micron All-Sky Survey
VLT   Very Large Telescope
VIMOS   Visible Multi Object Spectrograph
GBM   [Fermi] Gamma-ray Burst Monitor
XRT   [Swift] X-Ray Telescope
IPN   Inter-Planetary Network

References

1. Kouveliotou, C.; Meegan, C.A.; Fishman, G.J.; Bhat, N.P.; Briggs, M.S.; Koshut, T.M.; Paciesas, W.S.; Pendleton, G.N. Identification of Two Classes of Gamma-Ray Bursts. *Astrophys. J.* **1993**, *413*, L101. [CrossRef]
2. Bromberg, O.; Nakar, E.; Piran, T.; Sari, R. Short versus Long and Collapsars versus Non-collapsars: A Quantitative Classification of Gamma-Ray Bursts. *Astrophys. J.* **2013**, *764*, 179. [CrossRef]
3. Eichler, D.; Livio, M.; Piran, T.; Schramm, D.N. Nucleosynthesis, neutrino bursts and γ-rays from coalescing neutron stars. *Nature* **1989**, *340*, 126–128. [CrossRef]
4. Narayan, R.; Paczynski, B.; Piran, T. Gamma-Ray Bursts as the Death Throes of Massive Binary Stars. *Astrophys. J.* **1992**, *395*, L83. [CrossRef]
5. Mochkovitch, R.; Hernanz, M.; Isern, J.; Martin, X. Gamma-ray bursts as collimated jets from neutron star/black hole mergers. *Nature* **1993**, *361*, 236–238. [CrossRef]
6. Bogomazov, A.I.; Lipunov, V.M.; Tutukov, A.V. Evolution of close binaries and gamma-ray bursts. *Astron. Rep.* **2007**, *51*, 308–317. [CrossRef]
7. Roberts, L.F.; Kasen, D.; Lee, W.H.; Ramirez-Ruiz, E. Electromagnetic transients powered by nuclear decay in the tidal tails of coalescing compact binaries. *Astrophys. J. Lett.* **2011**, *736*, L21. [CrossRef]
8. Giacomazzo, B.; Perna, R.; Rezzolla, L.; Troja, E.; Lazzati, D. Compact binary progenitors of short gamma-ray bursts. *Astrophys. J. Lett.* **2012**, *762*, L18. [CrossRef]
9. Paschalidis, V. General relativistic simulations of compact binary mergers as engines for short gamma-ray bursts. *Class. Quantum Gravity* **2017**, *34*, 084002. [CrossRef]
10. Berger, E. Short-duration gamma-ray bursts. *Annu. Rev. Astron. Astrophys.* **2014**, *52*, 43–105. [CrossRef]
11. Fong, W.F.; Berger, E. The locations of short gamma-ray bursts as evidence for compact object binary progenitors. *Astrophys. J.* **2013**, *776*, 18. [CrossRef]
12. Tunnicliffe, R.L.; Levan, A.J.; Tanvir, N.R.; Rowlinson, A.; Perley, D.A.; Bloom, J.S.; Cenko, S.B.; O'Brien, P.T.; Cobb, B.E.; Wiersema, K.; et al. On the nature of the 'hostless' short GRBs. *Mon. Not. R. Astron. Soc.* **2013**, *437*, 1495–1510. [CrossRef]
13. Bray, J.C.; Eldridge, J.J. Neutron star kicks and their relationship to supernovae ejecta mass. *Mon. Not. R. Astron. Soc.* **2016**, *461*, 3747–3759. [CrossRef]
14. Tanvir, N.R.; Levan, A.J.; Fruchter, A.S.; Hjorth, J.; Hounsell, R.A.; Wiersema, K.; Tunnicliffe, R.L. A 'kilonova' associated with the short-duration γ-ray burst GRB 130603B. *Nature* **2013**, *500*, 547–549. [CrossRef] [PubMed]
15. Berger, E.; Fong, W.; Chornock, R. An r-process kilonova associated with the short-hard GRB 130603B. *Astrophys. J. Lett.* **2013**, *774*, L23. [CrossRef]
16. Jin, Z.P.; Li, X.; Cano, Z.; Covino, S.; Fan, Y.Z.; Wei, D.M. The Light Curve of the Macronova Associated With the Long–short Burst GRB 060614. *Astrophys. J. Lett.* **2015**, *811*, L22. [CrossRef]
17. Gompertz, B.P.; Levan, A.J.; Tanvir, N.R.; Hjorth, J.; Covino, S.; Evans, P.A.; Fruchter, A.S.; González-Fernández, C.; Jin, Z.P.; Lyman, J.D.; et al. The Diversity of Kilonova Emission in Short Gamma-Ray Bursts. *Astrophys. J.* **2018**, *860*, 62. [CrossRef]
18. Abbott, B.P.; Abbott, R.; Abbott, T.D.; Acernese, F.; Ackley, K.; Adams, C.; Adams, T.; Addesso, P.; Adhikari, R.X.; Adya, V.B.; et al. Gravitational Waves and Gamma-Rays from a Binary Neutron Star Merger: GW170817 and GRB 170817A. *Astrophys. J.* **2017**, *848*, L13. [CrossRef]

19. Beniamini, P.; Petropoulou, M.; Barniol Duran, R.; Giannios, D. A lesson from GW170817: most neutron star mergers result in tightly collimated successful GRB jets. *Mon. Not. R. Astron. Soc.* **2018**, 2945. [CrossRef]
20. Tanvir, N.R.; Chapman, R.; Levan, A.J.; Priddey, R.S. An origin in the local Universe for some short γ-ray bursts. *Nature* **2005**, *438*, 991. [CrossRef] [PubMed]
21. Chapman, R.; Priddey, R.S.; Tanvir, N.R. Short gamma-ray bursts from SGR giant flares and neutron star mergers: two populations are better than one. *Mon. Not. R. Astron. Soc.* **2009**, *395*, 1515–1522. [CrossRef]
22. Rowlinson, A.; Wiersema, K.; Levan, A.J.; Tanvir, N.R.; O'Brien, P.T.; Rol, E.; Hjorth, J.; Thöne, C.C.; de Ugarte Postigo, A.; Fynbo, J.P.U.; et al. Discovery of the afterglow and host galaxy of the low-redshift short GRB 080905A. *Mon. Not. R. Astron. Soc.* **2010**, *408*, 383–391. [CrossRef]
23. de Ugarte Postigo, A.; Castro-Tirado, A.J.; Guziy, S.; Gorosabel, J.; Jóhannesson, G.; Aloy, M.A.; McBreen, S.; Lamb, D.Q.; Benitez, N.; Jelínek, M.; et al. GRB 060121: Implications of a Short-/Intermediate-Duration γ-Ray Burst at High Redshift. *Astrophys. J. Lett.* **2006**, *648*, L83–L87. [CrossRef]
24. Selsing, J.; Krühler, T.; Malesani, D.; D'Avanzo, P.; Schulze, S.; Vergani, S.D.; Palmerio, J.; Japelj, J.; Milvang-Jensen, B.; Watson, D.; et al. The host galaxy of the short GRB 111117A at z= 2.211-Impact on the short GRB redshift distribution and progenitor channels. *Astron. Astrophys.* **2018**, *616*, A48. [CrossRef]
25. Lamb, G.P.; Kobayashi, S. Electromagnetic counterparts to structured jets from gravitational wave detected mergers. *Mon. Not. R. Astron. Soc.* **2017**, *472*, 4953–4964. [CrossRef]
26. Jin, Z.P.; Li, X.; Wang, H.; Wang, Y.Z.; He, H.N.; Yuan, Q.; Zhang, F.W.; Zou, Y.C.; Fan, Y.Z.; Wei, D.M. Short GRBs: Opening Angles, Local Neutron Star Merger Rate, and Off-axis Events for GRB/GW Association. *Astrophys. J.* **2018**, *857*, 128. [CrossRef]
27. Kathirgamaraju, A.; Barniol Duran, R.; Giannios, D. Off-axis short GRBs from structured jets as counterparts to GW events. *Mon. Not. R. Astron. Soc.* **2018**, *473*, L121–L125. [CrossRef]
28. Mooley, K.P.; Deller, A.T.; Gottlieb, O.; Nakar, E.; Hallinan, G.; Bourke, S.; Frail, D.A.; Horesh, A.; Corsi, A.; Hotokezaka, K. Superluminal motion of a relativistic jet in the neutron-star merger GW170817. *Nature* **2018**, *561*, 355–359. [CrossRef] [PubMed]
29. van Eerten, E.T.H.; Ryan, G.; Ricci, R.; Burgess, J.M.; Wieringa, M.; Piro, L.; Cenko, S.B.; Sakamoto, T. A year in the life of GW170817: the rise and fall of a structured jet from a binary neutron star merger. *ArXiv* **2018**, arXiv:1808.06617.
30. Lazzati, D.; López-Cámara, D.; Cantiello, M.; Morsony, B.J.; Perna, R.; Workman, J.C. Off-axis Prompt X-Ray Transients from the Cocoon of Short Gamma-Ray Bursts. *Astrophys. J.* **2017**, *848*, L6. [CrossRef]
31. Kasliwal, M.M.; Nakar, E.; Singer, L.P.; Kaplan, D.L.; Cook, D.O.; Van Sistine, A.; Lau, R.M.; Fremling, C.; Gottlieb, O.; Jencson, J.E.; et al. Illuminating gravitational waves: A concordant picture of photons from a neutron star merger. *Science* **2017**, *358*, 1559–1565. [CrossRef] [PubMed]
32. Gottlieb, O.; Nakar, E.; Piran, T.; Hotokezaka, K. A cocoon shock breakout as the origin of the γ-ray emission in GW170817. *Mon. Not. R. Astron. Soc.* **2018**, *479*, 588–600. [CrossRef]
33. Lamb, G.P.; Kobayashi, S. GRB 170817A as a jet counterpart to gravitational wave trigger GW 170817. *Mon. Not. R. Astron. Soc.* **2018**, *478*, 733–740. [CrossRef]
34. Troja, E.; Rosswog, S.; Gehrels, N. Precursors of Short Gamma-ray Bursts. *Astrophys. J.* **2010**, *723*, 1711–1717. [CrossRef]
35. Kochanek, C.S. Coalescing Binary Neutron Stars. *Astrophys. J.* **1992**, *398*, 234. [CrossRef]
36. Tsang, D. Shattering Flares during Close Encounters of Neutron Stars. *Astrophys. J.* **2013**, *777*, 103. [CrossRef]
37. Kyutoku, K.; Ioka, K.; Shibata, M. Ultrarelativistic electromagnetic counterpart to binary neutron star mergers. *Mon. Not. R. Astron. Soc.* **2014**, *437*, L6–L10. [CrossRef]
38. Metzger, B.D.; Zivancev, C. Pair fireball precursors of neutron star mergers. *Mon. Not. R. Astron. Soc.* **2016**, *461*, 4435–4440. [CrossRef]
39. Kouveliotou, C.; Strohmayer, T.; Hurley, K.; van Paradijs, J.; Finger, M.H.; Dieters, S.; Woods, P.; Thompson, C.; Duncan, R.C. Discovery of a Magnetar Associated with the Soft Gamma Repeater SGR 1900+14. *Astrophys. J. Lett.* **1999**, *510*, L115–L118. [CrossRef]
40. Palmer, D.M.; Barthelmy, S.; Gehrels, N.; Kippen, R.M.; Cayton, T.; Kouveliotou, C.; Eichler, D.; Wijers, R.A.M.J.; Woods, P.M.; Granot, J.; et al. A giant γ-ray flare from the magnetar SGR 1806 - 20. *Nature* **2005**, *434*, 1107–1109. [CrossRef] [PubMed]

41. Hurley, K.; Boggs, S.E.; Smith, D.M.; Duncan, R.C.; Lin, R.; Zoglauer, A.; Krucker, S.; Hurford, G.; Hudson, H.; Wigger, C.; et al. An exceptionally bright flare from SGR 1806-20 and the origins of short- duration γ-ray bursts. *Nature* **2005**, *434*, 1098–1103. [CrossRef] [PubMed]
42. Hurley, K. The short gamma-ray burst - SGR giant flare connection. *Adv. Space Res.* **2011**, *47*, 1337–1340. [CrossRef]
43. Bibby, J.L.; Crowther, P.A.; Furness, J.P.; Clark, J.S. A downward revision to the distance of the 1806-20 cluster and associated magnetar from Gemini Near-Infrared Spectroscopy. *Mon. Not. R. Astron. Soc.* **2008**, *386*, L23–L27. [CrossRef]
44. Cline, T.L.; Desai, U.D.; Pizzichini, G.; Teegarden, B.J.; Evans, W.D.; Klebesadel, R.W.; Laros, J.G.; Hurley, K.; Niel, M.; Vedrenne, G. Detection of a fast, intense and unusual gamma-ray transient. *Astrophys. J.* **1980**, *237*, L1–L5. [CrossRef]
45. Thompson, C.; Duncan, R.C. The Giant Flare of 1998 August 27 from SGR 1900+14. II. Radiative Mechanism and Physical Constraints on the Source. *Astrophys. J.* **2001**, *561*, 980–1005. [CrossRef]
46. Gehrels, N.; Chincarini, G.; Giommi, P.; Mason, K.O.; Nousek, J.A.; Wells, A.A.; White, N.E.; Barthelmy, S.D.; Burrows, D.N.; Cominsky, L.R.; et al. The Swift gamma-ray burst mission. *Astrophys. J.* **2004**, *611*, 1005. [CrossRef]
47. Barthelmy, S.D.; Barbier, L.M.; Cummings, J.R.; Fenimore, E.E.; Gehrels, N.; Hullinger, D.; Krimm, H.A.; Markwardt, C.B.; Palmer, D.M.; Parsons, A.; et al. The Burst Alert Telescope (BAT) on the SWIFT Midex mission. *Space Sci. Rev.* **2005**, *120*, 143–164. [CrossRef]
48. Zhang, B.; Fan, Y.Z.; Dyks, J.; Kobayashi, S.; Mészáros, P.; Burrows, D.N.; Nousek, J.A.; Gehrels, N. Physical Processes Shaping Gamma-Ray Burst X-Ray Afterglow Light Curves: Theoretical Implications from the Swift X-Ray Telescope Observations. *Astrophys. J.* **2006**, *642*, 354. [CrossRef]
49. Brown, P.J.; Holland, S.T.; Immler, S.; Milne, P.; Roming, P.W.A.; Gehrels, N.; Nousek, J.; Panagia, N.; Still, M.; Vanden Berk, D. Ultraviolet Light Curves of Supernovae with the Swift Ultraviolet/Optical Telescope. *Astrophys. J.* **2009**, *137*, 4517.
50. Tanvir, N.R.; Levan, A.J.; González-Fernández, C.; Korobkin, O.; Mandel, I.; Rosswog, S.; Hjorth, J.; D'Avanzo, P.; Fruchter, A.S.; Fryer, C.L.; et al. The Emergence of a Lanthanide-rich Kilonova Following the Merger of Two Neutron Stars. *Astrophys. J. Lett.* **2017**, *848*, L27. [CrossRef]
51. Levan, A.J.; Tanvir, N.R.; Jakobsson, P.; Chapman, R.; Hjorth, J.; Priddey, R.S.; Fynbo, J.P.U.; Hurley, K.; Jensen, B.L.; Johnson, R.; et al. On the nature of the short-duration GRB 050906. *Mon. Not. R. Astron. Soc.* **2008**, *384*, 541–547. [CrossRef]
52. Perley, D.A.; Bloom, J.S.; Modjaz, M.; Miller, A.A.; Shiode, J.; Brewer, J.; Starr, D.; Kennedy, R. GRB 070809: Putative Host Galaxy and Redshift. 2008. Available online: https://gcn.gsfc.nasa.gov/gcn3/7889.gcn3 (accessed on 29 November 2018).
53. O'Brien, P.T.; Tanvir, N.R. GRB 090417A: Nearby Galaxy Redshift. 2009. Available online: https://gcn.gsfc.nasa.gov/other/090417A.gcn3 (accessed on 29 November 2018).
54. Huchra, J.P.; Macri, L.M.; Masters, K.L.; Jarrett, T.H.; Berlind, P.; Calkins, M.; Crook, A.C.; Cutri, R.; Erdogdu, P.; Falco, E.; et al. The 2MASS Redshift Survey – Description and Data Release. *Astrophys. J. Suppl.* **2012**, *199*, 26. [CrossRef]
55. Makarov, D.; Prugniel, P.; Terekhova, N.; Courtois, H.; Vauglin, I. HyperLEDA. III. The catalogue of extragalactic distances. *Astron. Astrophys.* **2014**, *570*, A13. [CrossRef]
56. Helou, G.; Madore, B.F.; Schmitz, M.; Bicay, M.D.; Wu, X.; Bennett, J. The NASA/IPAC extragalactic database. In *Databases & On-Line Data in Astronomy*; Springer: Berlin, Germany, 1991; pp. 89–106.
57. Tully, R.B.; Courtois, H.M.; Sorce, J.G. Cosmicflows-3. *Astron. J.* **2016**, *152*, 50. [CrossRef]
58. Fong, W.; Berger, E.; Chornock, R.; Margutti, R.; Levan, A.J.; Tanvir, N.R.; Tunnicliffe, R.L.; Czekala, I.; Fox, D.B.; Perley, D.A.; et al. Demographics of the Galaxies Hosting Short-duration Gamma-Ray Bursts. *Astrophys. J.* **2013**, *769*, 56. [CrossRef]
59. Tanvir, N.R.; Malesani, D. GRB 111210A: SDSS Prior Imaging. 2011. Available online: https://gcn.gsfc.nasa.gov/gcn3/12661.gcn3 (accessed on 29 November 2018).
60. Levan, A.J.; Tanvir, N.R. GRB 130515A: FORS2 Spectroscopy of Candidate Counterpart. 2013. Available online: https://gcn.gsfc.nasa.gov/gcn3/14667.gcn3 (accessed on 29 November 2018).

61. Fong, W.; Berger, E.; Margutti, R.; Zauderer, B.A.; Troja, E.; Czekala, I.; Chornock, R.; Gehrels, N.; Sakamoto, T.; Fox, D.B.; et al. A Jet Break in the X-Ray Light Curve of Short GRB 111020A: Implications for Energetics and Rates. *Astrophys. J.* **2012**, *756*, 189. [CrossRef]
62. Cenko, S.B.; Cucchiara, A. GRB 130515A: Further Gemini Observations. 2013. Available online: https://gcn.gsfc.nasa.gov/gcn3/14670.gcn3 (accessed on 29 November 2018).
63. Connaughton, V.; Briggs, M.S.; Goldstein, A.; Meegan, C.A.; Paciesas, W.S.; Preece, R.D.; Wilson-Hodge, C.A.; Gibby, M.H.; Greiner, J.; Gruber, D.; Jenke, P.; Kippen, R.M.; et al. Localization of Gamma-Ray Bursts Using the Fermi Gamma-Ray Burst Monitor. *Astrophys. J. Suppl. Ser.* **2015**, *216*, 32. [CrossRef]
64. Briggs, M.S.; Pendleton, G.N.; Kippen, R.M.; Brainerd, J.J.; Hurley, K.; Connaughton, V.; Meegan, C.A. The error distribution of BATSE gamma-ray burst locations. *Astrophys. J. Suppl.* **1999**, *122*, 503. [CrossRef]
65. Karachentsev, I.D.; Makarov, D.I.; Kaisina, E.I. Updated Nearby Galaxy Catalog. *Astron. J.* **2013**, *145*, 101. [CrossRef]
66. Lazzati, D.; Ghirlanda, G.; Ghisellini, G. Soft gamma-ray repeater giant flares in the BATSE short gamma-ray burst catalogue: constraints from spectroscopy. *Mon. Not. R. Astron. Soc.* **2005**, *362*, L8–L12. [CrossRef]
67. Popov, S.B.; Stern, B.E. Soft gamma repeaters outside the Local Group. *Mon. Not. R. Astron. Soc.* **2006**, *365*, 885–890. [CrossRef]
68. Nakar, E.; Gal-Yam, A.; Piran, T.; Fox, D.B. The Distances of Short-Hard Gamma-Ray Bursts and the Soft Gamma-Ray Repeater Connection. *Astrophys. J.* **2006**, *640*, 849–853. [CrossRef]
69. Ofek, E.O. Soft Gamma-Ray Repeaters in Nearby Galaxies: Rate, Luminosity Function, and Fraction among Short Gamma-Ray Bursts. *Astrophys. J.* **2007**, *659*, 339–346. [CrossRef]
70. Tikhomirova, Y.Y.; Pozanenko, A.S.; Hurley, K.S. Search for nearby host galaxies of short gamma-ray bursts detected and well localized by BATSE/IPN. *Astron. Lett.* **2010**, *36*, 231–236. [CrossRef]
71. Svinkin, D.S.; Hurley, K.; Aptekar, R.L.; Golenetskii, S.V.; Frederiks, D.D. A search for giant flares from soft gamma-ray repeaters in nearby galaxies in the Konus-WIND short burst sample. *Mon. Not. R. Astron. Soc.* **2015**, *447*, 1028–1032. [CrossRef]
72. Evans, W.D.; Klebesadel, R.W.; Laros, J.G.; Cline, T.L.; Desai, U.D.; Teegarden, B.J.; Pizzichini, G.; Hurley, K.; Niel, M.; Vedrenne, G. Location of the gamma-ray transient event of 1979 March 5. *Astrophys. J.* **1980**, *237*, L7–L9. [CrossRef]
73. Cline, T.L.; Desai, U.D.; Teegarden, B.J.; Evans, W.D.; Klebesadel, R.W.; Laros, J.G.; Barat, C.; Hurley, K.; Niel, M.; Bedrenne, G.; et al. Precise source location of the anomalous 1979 March 5 gamma ray transient. *Astrophys. J.* **1981**, *255*, L45–L48. [CrossRef]
74. Perley, D.A.; Bloom, J.S. GRB 070201: Proximity of IPN Annulus to M31. 2007. Available online: https://gcn.gsfc.nasa.gov/gcn3/6091.gcn3 (accessed on 29 November 2018).
75. Hurley, K.; Rowlinson, A.; Bellm, E.; Perley, D.; Mitrofanov, I.G.; Golovin, D.V.; Kozyrev, A.S.; Litvak, M.L.; Sanin, A.B.; Boynton, W.; et al. A new analysis of the short-duration, hard-spectrum GRB 051103, a possible extragalactic soft gamma repeater giant flare. *Mon. Not. R. Astron. Soc.* **2010**, *403*, 342–352. [CrossRef]
76. Levan, A.J.; Tanvir, N.R.; Hjorth, J. Short GRB 150906B: Proximity to NGC 3313 Galaxy Group. 2015. Available online: https://gcn.gsfc.nasa.gov/gcn3/18263.gcn3 (accessed on 29 November 2018).
77. Abbott, B.; Abbott, R.; Adhikari, R.; Agresti, J.; Ajith, P.; Allen, B.; Amin, R.; Anderson, S.B.; Anderson, W.G.; Arain, M.; et al. Implications for the Origin of GRB 070201 from LIGO Observations. *Astrophys. J.* **2008**, *681*, 1419. [CrossRef]
78. Abadie, J.; Abbott, B.P.; Abbott, T.D.; Abbott, R.; Abernathy, M.; Adams, C.; Adhikari, R.; Affeldt, C.; Ajith, P.; Allen, B.; et al. Implications for the Origin of GRB 051103 from LIGO Observations. *Astrophys. J.* **2012**, *755*, 2. [CrossRef]
79. Abbott, B.P.; Abbott, R.; Abbott, T.D.; Abernathy, M.R.; Acernese, F.; Ackley, K.; Adams, C.; Adams, T.; Addesso, P.; Adhikari, R.X.; et al. Search for gravitational waves associated with gamma-ray bursts during the first advanced LIGO observing run and implications for the origin of GRB 150906B. *Astrophys. J.* **2017**, *841*, 89. [CrossRef]
80. Abbott, B.P.; Abbott, R.; Abbott, T.D.; Acernese, F.; Ackley, K.; Adams, C.; Adams, T.; Addesso, P.; Adhikari, R.X.; Adya, V.B.; et al. GW170817: Observation of Gravitational Waves from a Binary Neutron Star Inspiral. *Phys. Rev. Lett.* **2017**, *119*, 161101. [CrossRef] [PubMed]
81. Chruslinska, M.; Belczynski, K.; Klencki, J.; Benacquista, M. Double neutron stars: merger rates revisited. *Mon. Not. R. Astron. Soc.* **2018**, *474*, 2937–2958. [CrossRef]

82. Troja, E.; Lipunov, V.M.; Mundell, C.G.; Butler, N.R.; Watson, A.M.; Kobayashi, S.; Cenko, S.B.; Marshall, F.E.; Ricci, R.; Fruchter, A.; et al. Significant and variable linear polarization during the prompt optical flash of GRB 160625B. *Nature* **2017**, *547*, 425–427. [CrossRef] [PubMed]
83. Hallinan, G.; Corsi, A.; Mooley, K.P.; Hotokezaka, K.; Nakar, E.; Kasliwal, M.M.; Kaplan, D.L.; Frail, D.A.; Myers, S.T.; Murphy, T.; et al. A radio counterpart to a neutron star merger. *Science* **2017**, *358*, 1579–1583. [CrossRef] [PubMed]
84. Lyman, J.D.; Lamb, G.P.; Levan, A.J.; Mandel, I.; Tanvir, N.R.; Kobayashi, S.; Gompertz, B.; Hjorth, J.; Fruchter, A.S.; Kangas, T.; et al. The optical afterglow of the short gamma-ray burst associated with GW170817. *Nat. Astron.* **2018**, *2*, 751–754. [CrossRef]
85. Evans, P.A.; Cenko, S.B.; Kennea, J.A.; Emery, S.W.K.; Kuin, N.P.M.; Korobkin, O.; Wollaeger, R.T.; Fryer, C.L.; Madsen, K.K.; Harrison, F.A.; et al. Swift and NuSTAR observations of GW170817: Detection of a blue kilonova. *Science* **2017**, *358*, 1565–1570. [CrossRef] [PubMed]
86. Licquia, T.C.; Newman, J.A. Improved Estimates of the Milky Way's Stellar Mass and Star Formation Rate from Hierarchical Bayesian Meta-Analysis. *Astrophys. J.* **2015**, *806*, 96. [CrossRef]
87. Abbott, B.P.; Abbott, R.; Abbott, T.D.; Abernathy, M.R.; Acernese, F.; Ackley, K.; Adams, C.; Adams, T.; Addesso, P.; Adhikari, R.X.; et al. Prospects for observing and localizing gravitational-wave transients with Advanced LIGO, Advanced Virgo and KAGRA. *Living Rev. in Relativ.* **2018**, *21*, 3. [CrossRef] [PubMed]

© 2018 by the authors. Licensee MDPI, Basel, Switzerland. This article is an open access article distributed under the terms and conditions of the Creative Commons Attribution (CC BY) license (http://creativecommons.org/licenses/by/4.0/).

Article

The Host Galaxies of Short GRBs as Probes of Their Progenitor Properties

Massimiliano De Pasquale

Department of Astronomy and Space Sciences, Faculty of Science, Istanbul University, 34119 Istanbul, Turkey; m.depasquale@ucl.ac.uk

Received: 30 November 2018; Accepted: 22 January 2019; Published: 7 February 2019

Abstract: We present and discuss the properties of host galaxies of short Gamma-ray Burst (SGRBs). In particular, we examine those observations that contribute to the understanding of the progenitor systems of these explosions. Most SGRB hosts are found to be star forming objects, but an important fraction, ∼1/5, of all hosts are elliptical with negligible star formation. Short bursts often occur at very large off-sets from their hosts, in regions where there is little or no underlying host light. Such results have enabled the community to test and improve the models for the production of short GRBs. In particular, the data are in favour of the merger of compact object binaries, provided that the kick velocities from the birth site are a few tens of km/s, and merger times of ∼1 Gyr.

Keywords: short GRBs; galaxies; compact object mergers

1. Introduction

It was discovered in 1993 [1], thanks to the Burst And Transient Source Experiment (BATSE) instrument onboard the Compton Gamma-ray Observatory, that Gamma-ray Bursts (GRBs) are divided into two classes of objects that show different observational properties. In particular, there is a clear bimodal distribution in the parameter T_{90}, the time interval in which 90% of the counts of the prompt gamma-ray emission are collected, which shows two peaks at ≃0.3 and ≃25 s. Moreover, the spectra of the "short" GRBs are, on average, harder than the spectra of the "long" ones (LGRBs henceforth). While the exact values of peaks of the duration distributions and the divide depends on the observing instrument, and there is a certain degree of overlap in the distributions of T_{90} between the two classes, these early studies demonstrated that GRBs were not homogeneous. Differences emerged also when the first GRB afterglows were discovered and studied. Before the launch of the *Neil Gehrels Swift Observatory* mission [2], precise positions of SGRBs could not be obtained from their prompt emission (as was possible for LGRBs), which prevented the community from observing SGRB afterglows. However, in the rare cases in which a SGRB could be pinpointed with precision good enough to permit rapid follow up observations with narrow field instruments, afterglows were not seen in any electromagnetic band ([3–5] and references therein) All these pieces of information contributed to the idea that SGRB had different progenitors and "central engine" from those of long GRBs. In particular, the short duration indicated very compact progenitors and the lack of an afterglow pointed to a very thin circumburst medium (which would produce the observed afterglow emission). These characteristics pointed to the direction of mergers of binary neutron stars (NSs) or of NS–black hole (BH) systems [6–8]. To determine which physical models for short GRBs were viable, however, we still needed to find their afterglows. This way, we could constrain essential parts of the physics of these sources, such as the energy and the geometry of the ejecta. Furthermore, afterglows could point to the locations and the environments of the galactic systems that were the abodes of these explosions. The importance of positions and environs of cosmic sources cannot be overestimated: they indicate the nature of the progenitors. For example, LGRBs occur in star forming galaxies, in positions consistent with regions where new stars are born, while no LGRB has ever been associated with passive, early type galaxies (the possible exception of GRB 050219B [9], can be still explained as a chance superposition). This finding strongly supported the notion that the progenitor of LGRBs were massive stars, which did not live long and could not move much from their birth site. The situation changed with the discovery of SGRB afterglows, made possible thanks to *Swift*. Thanks to its very rapid repointing capabilities, *Swift* has been able to find the weak X-ray afterglows of short GRBs and thus determine the positions of these sources with an error of a few arcseconds; in many cases, *Swift* observations have also enabled the discovery of optical afterglows, which has led to positions of sub-arcsecond precision. Thanks to this accuracy, we have determined the host galaxies of tens of short GRBs, and have found the most likely hosts of many other events. In this review, we briefly summarize the results of systematic observations, and what they imply for the SRB progenitors.

2. Statistics of SGRB Host Galaxies

In the case of SGRBs with optical afterglows, the optical source enables us to pinpoint the GRB with a sub-arcsecond accuracy, and the bright galaxy in the immediate vicinity of the afterglow is generally regarded as the host of the event. However, as Reference [10] first found out, a few SGRB with optical afterglow occur in a relatively large region devoid of galaxies down to deep upper limits, typically of magnitude $m \gtrsim 25$–26. By means of probabilistic arguments, based on the size of such region, the density of field galaxies brighter than a certain magnitude, the afterglow position and the distance between a certain galaxy and the afterglow position, reference [11] derives the probability that a galaxy close by is not the host galaxy of the event (chance association, P_c). Reference [11] regards a galaxy as the host of the event only if the P_c is less than 0.05. This allows the recovery a few more host GRB (see below); however, there are still SGRBs for which the methodology mentioned above yields a large probability (larger than 5%, and sometimes as large as ∼50%) that surrounding galaxies are serendipitous objects, i.e., they are not the true host of the event. Furthermore, deep photometry with the Hubble Space Telescope and ground telescopes have not revealed nearby hosts for these events. A way to interpret these events is that one of the surrounding, visible galaxies is the host of the event, in the sense that the progenitor was born in it, but such a progenitor escaped the galaxy at high speed and reached a very large distance.

SGRB optical afterglows are, on average, one order of magnitude less luminous than the optical afterglows of LGRBs [12]. Out of the 68 SGRBs discovered by *Swift* up to May 2012, only 23 (≃1/3) have an optical afterglow. According to the prevailing model, the GRB afterglow emission is synchrotron radiation; the optical band is likely between the synchrotron peak frequency and cooling frequency. This is a regime in which the flux depends on both the density of the circumburst medium and the kinetic energy of the ejecta. If there is a relation between the density and the host galaxy type, optical afterglows might be biased against early type galaxies. Getting a picture as complete and unbiased as possible of the host galaxies of GRBs is critical for shedding light on the progenitors. In addition, the age of the stellar population tells us about the time it takes for the progenitor system to eventually produce a short GRB. For these reasons, reference [11] has also studied the SGRBs for which only an X-ray afterglow is known. The *Swift* X-ray telescope (XRT) can usually produce positions with an accuracy of ≃1.5″ at a 90% confidence level (CL). By examining the *Swift* archive of observations up to 2012, reference [11] finds a total of 36 objects (optical + X-ray identified) with $P_c \lesssim 0.05$. When breaking down this list on the basis of galaxy type (see their article and the references within), they find that 47% are late-type objects, 17% are early type galaxies, and 19% are "inconclusive", i.e., the data are not good enough to assess the kind of galaxy. Finally, 17% SGRB are "host-less", i.e., they cannot be associated with a galaxy with $P_c < 0.05$. When assigning the host-less GRBs to their most probable host galaxies, one recovers similar figures: ≃50%, ≃20%, and ≃30% in the same categories as above (see Figure 1)[1].

A Kolmogorov-Smirnov (KS) test does not show a correlation between the type of host galaxy of a short GRB and its duration [11]. This is of particular interest since, as mentioned in the Introduction, there is an overlap between the properties of SGRBs and LGRBs, and some events due to collapsars might have been erroneously included into the SGRB class. In principle, this could explain why one finds more star forming galaxies than elliptical objects among the hosts of SGRBs. However, this has been found not to be true. Reference [13] has shown that one can assign to a GRB event the probability that it is not to a collapsar, based on its duration. Reference [11] has attested that, even if we examine only SGRBs with very high probability (>0.9) of not being caused by a collapsar, we find a ratio of star forming galaxies to passive galaxies *even larger* than that of the "complete" sample.

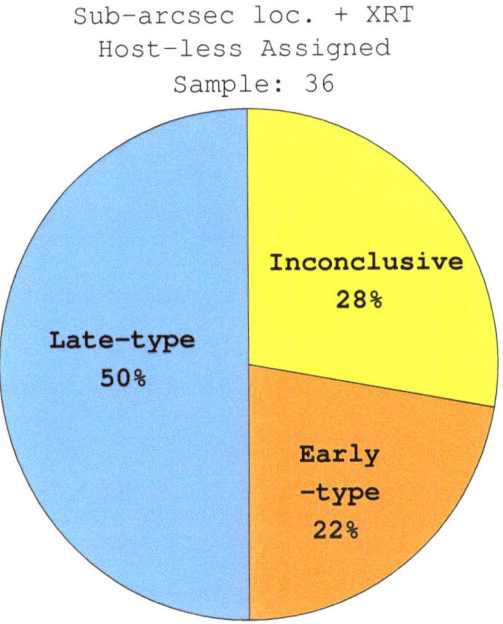

Figure 1. Pie diagram that shows the types of SGRB host galaxies. Source: Figure 13a in [11].

3. The Location of SGRB Afterglows in Their Host Galaxies

3.1. Off-Set Distribution

Reference [14] carried out a systematic study of the position of SGRBs with respect their host galaxies. Such an analysis was carried out by means of extensive Hubble Space Telescope imaging (see [14] for a detailed account of the procedure). It shows that the projected physical off-sets of the afterglows from the centres of the galaxies ranges from 0.5 to 75 kpc, while the median is 4.5 kpc. This is more than three times larger than the median off-sets for LGRBs (see Figure 2)[2], which is ≃1.3 kpc [15,16], but similar to the median off-sets of supernovae (SNe) Ia, which is ≃3 kpc. However, SGRBs can be placed further away from their hosts than SN Ia; the latter are not found beyond 20 kpc from their host, while 10% of SGRBs are. Reference [14] has also studied the distribution of

---

[1] http://iopscience.iop.org/article/10.1088/0004-637X/769/1/56/meta.
[2] http://iopscience.iop.org/article/10.1088/0004-637X/776/1/18/meta.

the "host-normalized" off-sets, i.e., the ratios between the off-set of the afterglow and the host galaxy effective radii r_e. This is of great importance because the large physical off-sets of the afterglow per se may not indicate that SGRBs occur preferentially outside their host galaxies, as predicted in some models of compact binary mergers, if the host galaxies themselves are large. It is found that the host-normalized off-set distribution ranges from 0 to almost 15, while the median is 1.5, with only 25% of events at $\lesssim 1\,r_e$. This contrasts with the host-normalized $r_e \simeq 1$ of LGRBs, core-collapse SNe, and even SN-Ia. The probability that LGRB and SGRB host-normalized off-sets are drawn from the same underlying population is only 3%, while the same probability for SGRBs and SN-Ia is less than 10^{-3}; these results indicate substantial differences between the progenitors of SGRBs and the other two classes of explosions.

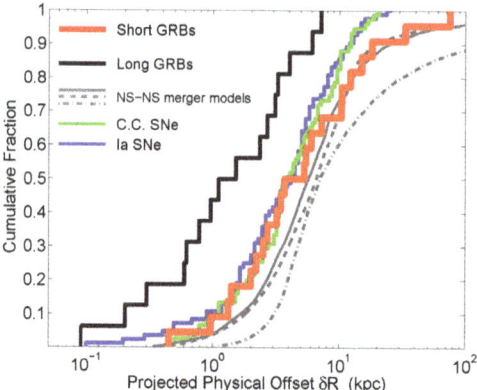

Figure 2. Cumulative distribution of projected physical off-sets for 22 short GRBs with sub-arcsecond positions (red; from [11]). Also shown are the cumulative distributions for long GRBs (black), core-collapse SNe (green), Type Ia SNe (blue); and predicted off-sets for NS–NS binaries (grey). Source: Figure 5 in [14].

3.2. Light Fraction Distribution

Reference [14] has also studied the so-called fractional flux for SGRB host, that is, the fraction of total host light in pixels fainter than or equal to the light in the pixel at the location of the transient [17]. Similar studies have already been performed in the case of hosts of core-collapse SNe, SN Ia, and LGRBs [16–18]. In their study of SGRB hosts, ref. [14] considered the fractional flux for UV light, with a rest-frame wavelength of less than 400 nm, and that for optical light, with a rest-frame wavelength larger than 400 nm. Naturally, this kind of study has to be restricted to those events that have an optical afterglow, giving a total of 20 SGRBs. Results are striking: 45% of SGRB are placed on the lowest level of optical brightness, i.e., 0, and 55% are on the lowest level of UV emission. The median fractional optical flux is $\simeq 0.15$, while the median fractional UV flux is $\simeq 0$.

By comparing the SGRB fractional UV flux distribution with the host's galaxy own UV light distribution by means of a KS test, there is a quite low probability $p = 0.01$ that the two distributions are drawn from the same intrinsic population, while this probability rises to 0.04 considering the optical light. These results show some similarities but differences as well with respect to those for SN Ia. About 34% of Ia SNe are placed on pixels that have zero UV flux; this percentage is lower but comparable to that for SGRBs. However, only 6% of SN Ia occur on regions of zero optical flux. The difference between SGRBs and LGRBs is even more obvious: the median fractional optical flux of LGRBs is 0.80, while 5% of LGRBs occur on pixels with zero flux. A comparable analysis for core collapse supernovae yields results similar to those of LGRBs.

4. Implications of Galaxy Demographics, Off-Sets and Fractional Flux Distributions

The ratio between star forming and passive hosts informs us about whether the SGRB rate is driven by star formation or stellar mass. A ratio 1:1 would indicate that the SGRB rate is driven by stellar mass alone, since the stellar mass in the two types of galaxies is roughly equal up to redshift $\simeq 1$; if instead the star formation rate influences the SGRB rate, one would expect a number of star forming, late galaxies larger than that of passive, early type galaxies. According to [11], the ratio of star forming galaxies to passive objects is $\simeq 2.5:1$ (see Figure 1), quite different from 1:1, and an F-test run shows that there is only a 4% probability that the observed distribution of host galaxy type can be derived from a population in which the intrinsic ratio is 1:1. Only if all the objects in the "inconclusive" category were early type would become the ratio 1:1, but this circumstance is rather unlikely. As a consequence, one can infer that the SGRB rate is proportional to both the star formation rate and the stellar mass.

The number of hosts falling in the category of star forming galaxies is larger than those belonging to the class of early, elliptical galaxies, in which few to no stars are being formed. However, the latter still represent an important fraction of the total. This behaviour resembles that of type Ia supernovae, which take place in both star forming galaxies and elliptical galaxies. This similarity already suggests that the progenitors of SGRBs, like those of SN Ia, are evolved systems.

It is also possible to argue that, if the delay time, i.e., the time from the birth of the progenitor system until the SGRB, were several billion years, then one should prevalently find early type galaxies associated with SGRBs at $z \sim 0$. This does not appear to be the case, suggesting instead that the delay times are shorter. This conclusion is in agreement with a systematic study of the spectral energy distributions (SED) of SGRB hosts [11,19], which shows stellar populations with ages $\lesssim$ a few Gyr. As a consequence, the average delay time for SGRB should be ~ 1 Gyr.

The analysis of afterglow off-sets strongly indicates that the SGRB do not trace star forming regions and they are weakly, if at all, correlated with the stellar mass of the host galaxy. This indicates that they travel large distances from their birth sites. Moreover, their progenitors must be very different from those of LGRBs, and to some extent not similar to that those of SN Ia. A further insight into the nature of the progenitors can be derived if we compare the ages of the host galaxy populations and the off-sets of the SGRBs. Assuming that the delay between the birth of the progenitor system and the merger is comparable to the average stellar population age, one can infer the typical velocity at which the system moved from the birth site, that is, the "kick" velocity. First, reference [14] notes that there does not appear to be a correlation between the normalized off-set and the mass of the host galaxy (see the previous section). This result suggests that the gravity of the host galaxy does not play an important role in determining the kick velocity. Secondly, the distribution of off-sets leads to a distribution of projected speeds of the SGRB progenitors between 2 and 150 km^{-1} s^{-1}. When considering the velocity dispersion of stars in the host, one finds projected kick velocities of $\simeq$20–140 km^{-1} s^{-1} with a median of $\simeq$60 km^{-1} s^{-1}. This velocity range is consistent with the interval of kick velocities derived for NS–NS binaries in the Milky Way. Again with the aim of shedding light on the progenitors of short GRBs, it is interesting to compare the off-set observations with the predictions of population synthesis models of NS–NS binary mergers [20]. As is shown in Figure 2, the observed distribution is broadly consistent with that predicted one, although the simulations predict a somewhat larger off-set. This might be explained by the fact that, in order to build the off-set distribution, one needs an optical afterglow; this excludes $\simeq$2/3 of SGRBs. In particular, an optical afterglow may be weaker and thus more likely to go undetected if the environment density is small (see above); this condition is more likely for GRBs occurring far away from the host galaxy. As a consequence, including events with X-ray only position may actually improve the agreement between the population synthesis and the real distribution of off-sets.

5. GW170817/GRB 170817A in the Context of SGRB Host Galaxies

On 17 August 2017, the SGRB 170817A was detected by *Fermi* and *INTEGRAL*, only two seconds after the Gravitational Wave signal [21] which flagged the coalescence of a binary neutron star system was detected by aLIGO. Subsequent electromagnetic follow up observations showed a new optical transient where *Fermi*, *INTEGRAL* and the aLIGO error regions overlapped. These observations proved that the SGRB and the new electromagnetic source were indeed the outcome of the merger observed in the GW channel.

The host galaxy of the GRB 170817A/GW 170817 event was NGC 4993, at $\simeq$40 Mpc from Earth. The property of this galaxy and the position of the source with respect to the galaxy body are consistent with those illustrated above. NGC 4993 is morphologically a lenticular galaxy, dominated by its bulge, although it also appears to be slightly distorted. The effective light radius of this object is $r_e \simeq 3$ kpc in the optical band, while it is $\simeq$2 kpc for near-infrared observations [22]. The transient occurred at a projected distance of $\simeq$2 kpc from the galaxy centre. These figures yield a normalized off-set of $\simeq$0.7. The star formation rate of the galaxy is negligible, estimated to be less than a few thousandths of $M_\odot$ yr^{-1}. According to [23], however, $\simeq$20 percent of NGC 4993 stellar population is $\simeq$1 Gyr old, while the rest is older than 5 Gyr. The fact that the merger occurred close to the centre of the host galaxy, together with the old age of stellar populations, *suggests* a relatively small kick velocity, although statistical and dynamical considerations indicate that it might have been as large as a few hundreds of km s^{-1} [23]. At the position of GRB 170817A, no bright source is detected; the fractional flux is $\simeq$0.6. This value is at the higher end of those seen in cosmological SGRBs (see previous sections). However, reference [23] points out that we may not detect the light from the faintest regions of the host galaxies of cosmological SGRBs, while the host galaxy of SGRB 170817A is located much closer to us.

6. Conclusions

About 50% or more of GRB host galaxies are late-type, star forming systems; only $\simeq$20% are definitely early-type, elliptical, passive objects. The remaining fraction is made of hosts we cannot identify or understand the features. These figures strongly indicate that the amount of star formation plays an important role in determining the SGRB rate, together with the stellar mass of the host. However, some SGRB hosts do not have star formation, and SGRBs often occur in the lowest UV luminosity pixel of their hosts. These facts indicate that SGRBs are not connected with *recent* star formation. Indeed, the age of stellar populations in the host galaxies is $\lesssim$ a few Gyr, suggesting that the timescale between the creation of the SGRB progenitor and the SGRB event itself is of the order of 1 Gyr. The median off-set of SGRBs is 4.5 kpc (about three times that of LGRBs) and, together with the typical stellar population age, points out typical projected speeds of $\simeq$60 km^{-1} s^{-1} for the progenitors from their birth site to the final explosion. All the results described above indicate, or at least are consistent with, the scenario in which SGRB progenitors are binary systems made of two compact objects, either two neutron stars or a neutron star and a black hole.

Funding: This research was funded by the Research Fund of the University of Istanbul project number 30901.

Acknowledgments: M.D.P. thanks Alice Breeveld for her help with the manuscript, and Wen-Fai Fong for granting the right to re-use material in her articles. Figures 1 and 2 are © AAS, reproduced with permission.

Conflicts of Interest: The author declares no conflict of interest.

References

1. Kouveliotou, C.; Meegan, C.A.; Fishman, G.J.; Bhat, N.P.; Briggs, M.S.; Koshut, T.M.; Paciesas, W.S.; Pendleton, G.N. Identification of two classes of gamma-ray bursts. *Astrophys. J.* **1993**, *413*, L101–L104. [CrossRef]
2. Gehrels, N.; Chincarini, G.; Giommi, P.; Mason, K.O.; Nousek, J.A.; Wells, A.A.; White, N.E.; Barthelmy, S.D.; Burrows, D.N.; Cominsky, L.R.; et al. The Swift Gamma-Ray Burst Mission. *Astrophys. J.* **2004**, *611*, 1005–1020. [CrossRef]

3. Klotz, A.; Boer, M.; Atteia, J.L. Observational constraints on the afterglow of GRB 020531. *Astron. Astrophys.* **2003**, *404*, 815–818. [CrossRef]
4. Hurley, K.; Berger, E.; Castro-Tirado, A.; Cerón, J.C.; Cline, T.; Feroci, M.; Frail, D.A.; Frontera, F.; Masetti, N.; Guidorzi, C.; et al. Afterglow Upper Limits for Four Short-Duration, Hard Spectrum Gamma-Ray Bursts. *Astrophys. J.* **2002**, *567*, 447–453. [CrossRef]
5. Gorosabel, J.; Fynbo, J.U.; Hjorth, J.; Wolf, C.; Andersen, M.I.; Pedersen, H.; Christensen, L.; Jensen, B.L.; Møller, P.; Afonso, J.; et al. Strategies for prompt searches for GRB afterglows: The discovery of the GRB 001011 optical/near-infrared counterpart using colour-colour selection. *Astron. Astrophys.* **2002**, *384*, 11–23. [CrossRef]
6. Narayan, R.; Paczyński, B.; Piran, T. Gamma-ray bursts as the death throes of massive binary stars. *Astrophys. J.* **1992**, *395*, L83–L86. [CrossRef]
7. Katz, J.L.; Canel, L.M. The Long and the Short of Gamma-Ray Bursts. *Astrophys. J.* **1996**, *471*, 915. [CrossRef]
8. Narayan, R.; Piran, T.; Kumar, P. Accretion Models of Gamma-Ray Bursts. *Astrophys. J.* **2001**, *557*, 949–957. [CrossRef]
9. Rossi, A.; Piranomonte, S.; Savaglio, S.; Palazzi, E.; Michałowski, M.J.; Klose, S.; Hunt, L.K.; Amati, L.; Elliott, J.; Greiner, J.; et al. A quiescent galaxy at the position of the long GRB 050219A. *Astron. Astrophys.* **2014**, *572*, A47. [CrossRef]
10. Fong, W.; Berger, E.; Fox, D.B. Hubble Space Telescope Observations of Short Gamma-Ray Burst Host Galaxies: Morphologies, Offsets, and Local Environments. *Astrophys. J.* **2010**, *708*, 9. [CrossRef]
11. Fong, W.; Berger, E.; Chornock, R.; Margutti, R.; Levan, A.J.; Tanvir, N.R.; Tunnicliffe, R.L.; Czekala, I.; Fox, D.B.; Perley, D.A.; et al. Demographics of the Galaxies Hosting Short-duration Gamma-Ray Bursts. *Astrophys. J.* **2013**, *769*, 56. [CrossRef]
12. Kann, D.A.; Klose, S.; Zhang, B.; Covino, S.; Butler, N.R.; Malesani, D.; Nakar, E.; Wilson, A.C.; Antonelli, L.A.; Chincarini, G.; et al. The Afterglows of Swift-era Gamma-Ray Bursts. II. Type I GRB versus Type II GRB Optical Afterglows. *Astrophys. J.* **2011**, *734*, 96. [CrossRef]
13. Bromberg, O.; Nakar, E.; Piran, T. An Observational Imprint of the Collapsar Model of Long Gamma-Ray Bursts. *Astrophys. J.* **2012**, *749*, 110. [CrossRef]
14. Fong, W.; Berger, E. The Locations of Short Gamma-Ray Bursts as Evidence for Compact Object Binary Progenitors. *Astrophys. J.* **2013**, *776*, 18. [CrossRef]
15. Bloom, J.S.; Kulkarni, S.R.; Djorgovski, S.G. The Observed Offset Distribution of Gamma-Ray Bursts from Their Host Galaxies: A Robust Clue to the Nature of the Progenitors. *Astron. J.* **2002**, *123*, 1111. [CrossRef]
16. Lyman, J.D.; Levan, A.J.; Tanvir, N.R.; Fynbo, J.P.U.; McGuire, J.T.W.; Perley, D.A.; Angus, C.R.; Bloom, J.S.; Conselice, C.J.; Fruchter, A.S.; et al. The host galaxies and explosion sites of long-duration gamma ray bursts: Hubble Space Telescope near-infrared imaging. *Mon. Not. R. Astron. Soc.* **2017**, *467*, 1795–1817. [CrossRef]
17. Fruchter, A.S.; Levan, A.J.; Strolger, L.; Vreeswijk, P.M.; Thorsett, S.E.; Bersier, D.; Burud, I.; Cerón, J.C.; Castro-Tirado, A.J.; Conselice, C.; et al. Long gamma-ray bursts and core-collapse supernovae have different environments. *Nature* **2006**, *441*, 463–468. [CrossRef] [PubMed]
18. Svensson, K.M.; Levan, A.J.; Tanvir, N.R.; Fruchter, A.S.; Strolger, L.G. The host galaxies of core-collapse supernovae and gamma-ray bursts. *Mon. Not. R. Astron. Soc.* **2010**, *405*, 57–76. [CrossRef]
19. Leibler, C.N.; Berger, E. The Stellar Ages and Masses of Short Gamma-ray Burst Host Galaxies: Investigating the Progenitor Delay Time Distribution and the Role of Mass and Star Formation in the Short Gamma-ray Burst Rate. *Astrophys. J.* **2010**, *725*, 1202. [CrossRef]
20. Belczynski, K.; Perna, R.; Bulik, T.; Kalogera, V.; Ivanova, N.; Lamb, D.Q. A Study of Compact Object Mergers as Short Gamma-Ray Burst Progenitors. *Astrophys. J.* **2006**, *648*, 1110. [CrossRef]
21. Abbott, B.P.; Abbott, R.; Adhikari, R.X.; Ananyeva, A.; Anderson, S.B.; Appert, S.; Arai, K.; Araya, M.C.; Barayoga, J.C.; Barish, B.C.; et al. Multi-messenger Observations of a Binary Neutron Star Merger. *Astrophys. J.* **2017**, *848*, 12. [CrossRef]

22. Im, M.; Yoon, Y.; Lee, S.K.J.; Lee, H.M.; Kim, J.; Lee, C.U.; Kim, S.L.; Troja, E.; Choi, C.; Lim, G.; et al. Distance and Properties of NGC 4993 as the Host Galaxy of the Gravitational-wave Source GW170817. *Astrophys. J.* **2017**, *849*, 16. [CrossRef]
23. Levan, A.J.; Lyman, J.D.; Tanvir, N.R.; Hjorth, J.; Mandel, I.; Stanway, E.R.; Steeghs, D.; Fruchter, A.S.; Troja, E.; Schrøder, S.L.; et al. The Environment of the Binary Neutron Star Merger GW170817. *Astrophys. J.* **2017**, *848*, 28. [CrossRef]

© 2019 by the author. Licensee MDPI, Basel, Switzerland. This article is an open access article distributed under the terms and conditions of the Creative Commons Attribution (CC BY) license (http://creativecommons.org/licenses/by/4.0/).

Article

Investigation of Similarity in the Spectra between Short- and Long-Duration Gamma-ray Bursts

Takanori Sakamoto *, Yuuki Yoshida and Motoko Serino

Department of Physics andMathematics, College of Science and Engineering, Aoyama Gakuin University, 5-10-1 Fuchinobe, Chuo-ku, Sagamihara-shi, Kanagawa 252-5258, Japan; yusakamoto.lab.0927@gmail.com (Y.Y.); serino@phys.aoyama.ac.jp (M.S.)
* Correspondence: tsakamoto@phys.aoyama.ac.jp; Tel.: +81-42-759-6275

Received: 8 August 2018; Accepted: 26 September 2018; Published: 3 October 2018

Abstract: We investigated the spectral properties of the prompt emission for short- and long-duration gamma-ray bursts (GRBs) using the *Fermi* Gamma-ray Burst Monitor data. In particular, we focused on comparing the spectral properties of short GRBs and the initial 2 s of long GRBs, motivated by the previous study of Ghirlanda et al. (2009). We confirmed the similarity in the low energy photon index α between short GRBs and the initial 2 s of long GRBs. Since about a quarter of our spectra of both short GRBs and the initial 2 s of long GRBs show α to be shallower than $-2/3$, it is difficult to understand in the context standard synchrotron emission.

Keywords: gamma-ray burst; prompt emission; spectrum

1. Introduction

The origin of short-duration gamma-ray bursts (hereafter short GRBs) is receiving great attention in the field of astrophysics. During the second Laser Interferometer Gravitational wave Observatory and Virgo observation run in 2017, the first gravitational wave event from the merging neutron stars, GW 170817, was observed [1]. The *Fermi Gamma-ray Space Telescope* (*Fermi*) Gamma-ray Burst Monitor (GBM) and the Anti-Coincidence Shield (ACS) of the SPI spectrometer on board the *INTernational Gamma-ray Astrophysics Laboratory* (*INTEGRAL*) detected a possible short GRB 170817A about 1.7 s after the merger time [2]. GRB 170817A has the t_{90} duration of 2.0 ± 0.5 s and the total radiation energy was 3–4 orders of magnitude lower than that of the typical short GRBs [3]. The origin of the weakness of GRB 170817A is still unclear. However, the recent X-ray e.g., [4,5] and radio e.g., [6] observations at the late phase suggest that the weak emission is consistent with an off-axis viewing effect (a weakened relativistic beaming effect; [7,8]) of a typical short GRB jet.

One of the well-known properties of short GRBs is the hardness of their spectra in the prompt emission. A short GRB tends to have a harder spectrum than a long duration GRB (hereafter long GRB) e.g., [9]. Ghirlanda et al. [10] show that the hardness of short GRBs is driven by a shallower (harder) low energy photon index α compared to that of long GRBs rather than the peak energy in the νF_ν spectrum E_{peak} by the *Compton Gamma-ray Observatory* Burst And Transient Source Experiment (BATSE) sample. Further investigation with the much larger GRB sample confirmed this spectral difference between short GRBs and long GRBs [11]. Ghirlanda et al. [11] also pointed out that a spectrum of the initial part (the first 1–2 s of the emission) of a long GRB shows a similarity to a short GRB. Both α and E_{peak} of short GRBs and the initial part of long GRBs have statistically similar distributions. This result indicates that a similar radiation process is involved between short and long GRBs.

The spectral evolution between long GRBs and short GRBs shows a similar behavior. Hakkila & Preece [12] resolved individual pulses during the prompt emission, and demonstrated the similarities of the pulse characteristics between long GRBs and short GRBs. They also suggested that the general spectral evolution seen over the burst episode can be understood by the hard-to-soft evolution of the individual pulse. According to Ghirlanda et al. [13], E_{peak} of time-resolved spectra of both long and short GRBs shows a positive correlation to the peak fluxes (or luminosity). Those previous works shed light on the similarity in the radiation process between long and short GRBs despite the different progenitors.

In this paper, we report the comparison of the spectral properties of a prompt emission between short and long GRBs by *Fermi* GBM data. The *Fermi* GBM possesses a large GRB sample with a good spectral coverage to derive the spectral parameters of a prompt emission for both short and long GRBs. It is worth investigating the similarity in the spectral properties between short and long GRBs reported by the BATSE sample using data of a different GRB instrument. In Section 2, we describe the investigation of a possible systematic error in the *Fermi* GBM data, the analysis method and the sample selection. We report the result of the comparisons of the prompt spectral parameters among short GRBs, long GRBs and the initial 2 s of long GRBs in Section 3. We discuss and summarize our results in Section 4. The quoted errors are at the 90% confidence level.

2. Analysis

The HEASOFT version 6.21 and the *Fermi* Science tool version v10r0p5 are used throughout the analysis. The spectrum of the *Neil Gehrels Swift Observatory* [14] Burst Alert Telescope [15] data is generated following the BAT analysis thread.[1] The spectrum of *Fermi* GBM data is extracted from the Time-Tagged Event (TTE) data [16] using gtbin. The background spectrum is selected as a pre-burst interval in a duration that is 3–5-times longer than a source time interval depending on the stability of the background. The energy range of the spectral analysis is 15–150 keV for the Swift/BAT data. The energy range to be used in the spectral analysis of the *Fermi* GBM data is investigated in the following section. The energy response function of *Fermi* GRB is downloaded from the GBM triggered data archive[2] for each GRB. The XSPEC version 12.9.1 software package was used for fitting the spectral data.

2.1. Identifying the Spectral Energy Range of the Fermi GBM Data

First, we investigate the energy range of the spectral data of the *Fermi* GBM by performing the joint spectral fit to the simultaneously detected bright GRBs with the *Swift* BAT. The *Swift* BAT has regularly performed the spectral calibration, collecting the Crab nebula data at the specific incident angles [17–19]. Furthermore, the spectral cross-calibration has been performed with the Konus–Wind and the Suzaku/WAM using the simultaneously detected bright GRBs [20]. Therefore, the systematic errors in the energy response function of the *Swift* BAT are well understood.

The joint spectral analyses of 37 simultaneously detected GRBs by the *Fermi* GBM and the *Swift* BAT were conducted. We used the standard energy range between 7 keV and 1 MeV for the *Fermi* GBM NaI instrument and between 150 keV and 40 MeV for the BGO instrument. Figure 1 shows an example of GRB 170705A. As can be seen in the left panel of Figure 1, there is a noticeable residual from the best fit model at the spectral bins between 7 keV and 30 keV in the *Fermi* GBM NaI data. The reduced χ^2 of the fit is 1.561 in 102 degree of freedom. By ignoring the spectral bins below 30 keV in the *Fermi* GBM NaI data, the fit was significantly improved with the reduced χ^2 of 1.067 in 81 degrees of freedom (right panel of Figure 1). We systematically investigated all 37 GRBs and confirmed that the reasonable joint fit was achieved using above 30 keV for the *Fermi* GBM NaI data. Therefore, based on this study, we decided to use the energy range between 30 keV and 1 MeV for the *Fermi* GBM NaI data. The standard energy range between 150 keV and 40 MeV is used for the *Fermi* GBM BGO data.

---

[1] https://swift.gsfc.nasa.gov/analysis/threads/batspectrumthread.html.
[2] https://heasarc.gsfc.nasa.gov/FTP/fermi/data/gbm/triggers/.

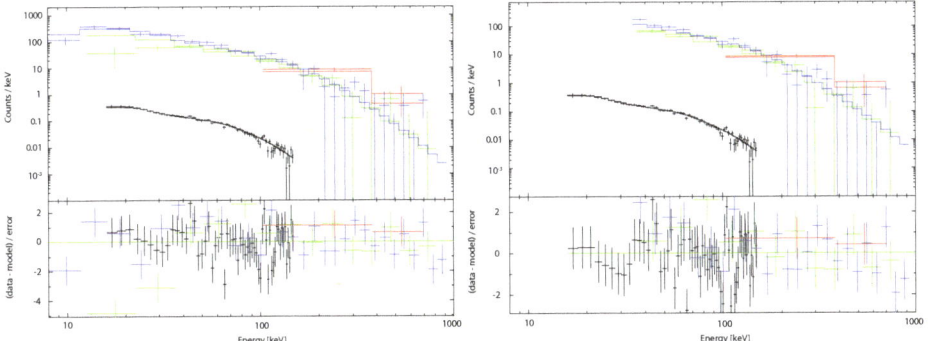

Figure 1. Joint spectral analysis of the *Swift* BAT (black) and the *Fermi* GBM NaI (green and blue) and BGO (red) for GRB 170705A (left: using the *Fermi* GBM NaI data from 7 keV to 1 MeV; right: using the *Fermi* GBM NaI data from 30 keV to 1 MeV).

2.2. Preparation of the Data

We selected 60 short GRBs and 58 long GRBs detected by the *Fermi* GBM between 2008 and 2017. The selection criteria of long GRBs are that the peak photon flux in 64 ms reported in the *Fermi* GBM Burst Catalog [21–23] is >7.1 ph cm^{-2} s^{-1} and the derived spectral parameters are well constrained. We selected all short GRBs during this period with well constrained spectral parameters. The foreground spectral files are generated using gtbin specifying the t_{90} interval for the time-averaged spectra of short and long GRBs. For the spectrum of the initial 2 s of long GRBs, we specified the 2 s window from the trigger time of the *Fermi* GBM. The background spectral files are generated using gtbin, specifying a pre-burst interval which is 3–5 times longer than a foreground interval. The data of two triggered NaI detectors and one BGO detector are used in the analysis. The data and the energy response files are downloaded from the *Fermi* GBM public data archive available through the *Fermi* Science Data Center.[3] The spectral model used in the fit is a cutoff power-law (CPL) model:

$$f(E) = K_{50} \left(\frac{E}{50 \text{ keV}}\right)^{\alpha} \exp\left(\frac{-E(2+\alpha)}{E_{\text{peak}}}\right)$$

where α is the low energy photon index, E_{peak} is the peak energy in the νF_ν spectrum and K_{50} is the normalization at 50 keV in units of photons cm^{-2} s^{-1} keV^{-1}.

3. Results

Tables 1–3 summarize our GRB samples and derived spectral parameters based on a CPL model fit. Figure 2 shows the distribution of E_{peak} and α between short GRBs and long GRBs (left panel), and the short GRBs and the initial 2 s of long GRBs (right panel). The distribution of α for long GRBs tends to overlap at a steeper region (small α) of short GRBs, whereas E_{peak} distributes to a higher energy for short GRBs compared to that of long GRBs. On the other hand, the difference in α becomes less evident between short GRBs and the initial 2 s of long GRBs.

---

[3] https://heasarc.gsfc.nasa.gov/FTP/fermi/data/gbm/bursts/.

Table 1. Spectral parameters of short GRBs.

GRB	α	E_{peak} [keV]
GRB081209981	-0.48 ± 0.40	848 ± 479
GRB081216531	-0.77 ± 0.18	791 ± 266
GRB081223419	-1.11 ± 0.36	301 ± 127
GRB081226509	0.47 ± 1.50	269 ± 133
GRB090228204	-0.49 ± 0.18	813 ± 194
GRB090305052	0.01 ± 0.29	611 ± 121
GRB090308734	-0.45 ± 0.22	519 ± 98
GRB090617208	-0.24 ± 0.64	823 ± 623
GRB090907808	0.50 ± 0.58	339 ± 59
GRB091126333	-0.39 ± 2.24	123 ± 63
GRB100206563	0.10 ± 0.77	349 ± 110
GRB100525744	-0.78 ± 0.81	540 ± 367
GRB100625773	-0.64 ± 0.49	504 ± 251
GRB100629801	-0.94 ± 0.31	233 ± 50
GRB100811108	0.31 ± 0.37	664 ± 141
GRB101031625	0.34 ± 1.22	292 ± 146
GRB101216721	-0.49 ± 0.25	140 ± 10
GRB110526715	-0.60 ± 0.36	422 ± 146
GRB110705151	-0.06 ± 0.27	818 ± 229
GRB111103948	0.35 ± 0.99	637 ± 330
GRB111222619	-0.48 ± 0.29	1147 ± 550
GRB120323507	-1.40 ± 0.38	599 ± 578
GRB120603439	-0.62 ± 0.50	444 ± 225
GRB120811014	0.10 ± 0.39	696 ± 178
GRB120817168	-0.40 ± 0.44	763 ± 448
GRB120830297	0.14 ± 0.25	737 ± 132
GRB121012724	-0.30 ± 0.45	393 ± 120
GRB130204484	1.52 ± 3.09	227 ± 152
GRB130307126	-0.24 ± 0.82	461 ± 286
GRB130628860	-0.64 ± 0.48	561 ± 331
GRB130701761	-0.56 ± 0.15	1551 ± 619
GRB130912358	-0.46 ± 0.40	409 ± 129
GRB131126163	1.23 ± 0.83	474 ± 92
GRB131217108	0.20 ± 0.74	697 ± 495
GRB140105065	-0.79 ± 0.59	316 ± 212
GRB140209313	-0.89 ± 0.08	210 ± 10
GRB140511095	0.18 ± 1.98	153 ± 76
GRB140605377	0.22 ± 1.00	326 ± 114
GRB140626843	-1.24 ± 0.52	108 ± 21
GRB140807500	-1.01 ± 0.24	762 ± 447
GRB140901821	-0.16 ± 0.33	1033 ± 420
GRB141011282	-0.32 ± 0.57	653 ± 256
GRB141105406	-0.79 ± 0.33	656 ± 341
GRB150118927	-0.98 ± 0.31	551 ± 323
GRB150506630	-0.44 ± 0.51	599 ± 299
GRB150604434	-0.72 ± 0.51	484 ± 295
GRB150811849	-0.09 ± 0.22	899 ± 205
GRB150819440	-1.28 ± 0.09	835 ± 292
GRB151231568	-0.95 ± 0.30	525 ± 246
GRB160406503	0.58 ± 1.22	323 ± 130
GRB160804180	-0.39 ± 0.38	555 ± 245
GRB160806584	-1.44 ± 0.27	149 ± 29
GRB160820496	-0.74 ± 0.56	412 ± 220
GRB160821937	-0.34 ± 2.04	101 ± 37
GRB160822672	-1.06 ± 0.80	252 ± 209
GRB170121133	1.82 ± 4.30	115 ± 46
GRB170127634	-0.67 ± 0.73	417 ± 290
GRB170206453	-0.67 ± 0.09	418 ± 36
GRB170305256	-0.54 ± 0.35	242 ± 50
GRB170325331	-0.65 ± 1.21	202 ± 153

Table 2. Spectral parameters of long GRBs.

GRB	α	E_{peak} [keV]
GRB081009140	−1.37 ± 0.14	36 ± 4
GRB081215784	−0.77 ± 0.02	566 ± 17
GRB090424592	−1.16 ± 0.04	190 ± 4
GRB090719063	−0.72 ± 0.04	277 ± 7
GRB090804940	−0.43 ± 0.10	103 ± 1
GRB091127976	−1.98 ± 0.00	9 ± 1
GRB100131730	−1.17 ± 0.08	220 ± 15
GRB100324172	−0.61 ± 0.05	483 ± 25
GRB100722096	−1.88 ± 0.11	31 ± 19
GRB100829876	−1.23 ± 0.08	225 ± 20
GRB100910818	−1.02 ± 0.09	179 ± 10
GRB101208498	−1.44 ± 0.15	115 ± 10
GRB110817191	−0.97 ± 0.07	222 ± 11
GRB110921912	−1.02 ± 0.05	643 ± 76
GRB111220486	−1.09 ± 0.06	334 ± 28
GRB120129580	−0.95 ± 0.03	347 ± 9
GRB120204054	−1.16 ± 0.04	198 ± 6
GRB120217904	−1.19 ± 0.09	319 ± 42
GRB120328268	−1.16 ± 0.03	345 ± 18
GRB120426090	−0.99 ± 0.05	148 ± 2
GRB120728434	−1.41 ± 0.05	96 ± 2
GRB120801920	−0.19 ± 0.50	440 ± 105
GRB130121835	−1.02 ± 0.30	235 ± 50
GRB130228212	−1.53 ± 0.10	268 ± 48
GRB130306991	−0.30 ± 0.84	170 ± 20
GRB130425327	−1.21 ± 0.13	252 ± 27
GRB130502327	−0.92 ± 0.07	645 ± 89
GRB130815660	−1.71 ± 0.15	134 ± 28
GRB130821674	−1.23 ± 0.07	493 ± 80
GRB131108862	−0.90 ± 0.06	432 ± 29
GRB131214705	−1.57 ± 0.07	107 ± 5
GRB131229277	−0.89 ± 0.08	360 ± 30
GRB140213807	−1.70 ± 0.06	106 ± 7
GRB140523129	−1.08 ± 0.03	293 ± 10
GRB140621827	−0.71 ± 0.14	571 ± 119
GRB140801792	−0.33 ± 0.09	125 ± 2
GRB141222298	−1.45 ± 0.07	1275 ± 796
GRB150330828	−1.10 ± 0.06	362 ± 30
GRB150403913	−0.99 ± 0.03	567 ± 29
GRB150426594	−0.74 ± 0.88	112 ± 24
GRB151227072	−1.06 ± 0.12	169 ± 10
GRB151227218	−1.45 ± 0.05	498 ± 93
GRB151231443	−0.91 ± 0.13	209 ± 12
GRB160113398	0.25 ± 1.91	140 ± 45
GRB160516237	−1.60 ± 0.46	77 ± 39
GRB160521385	−0.87 ± 0.05	205 ± 5
GRB160724444	−1.28 ± 0.09	200 ± 18
GRB16080225	−0.79 ± 0.03	347 ± 10
GRB160816730	−0.78 ± 0.04	264 ± 8
GRB160910722	−0.96 ± 0.02	457 ± 14
GRB161218356	−0.69 ± 0.03	245 ± 4
GRB170207906	0.00 ± 0.18	384 ± 29
GRB170511249	−1.38 ± 0.10	116 ± 6
GRB170522657	−0.61 ± 0.05	387 ± 15
GRB170626401	−1.26 ± 0.11	96 ± 4
GRB170802638	−0.44 ± 0.65	316 ± 132
GRB170826819	−0.88 ± 0.05	422 ± 24
GRB171120556	−1.23 ± 0.15	202 ± 27
GRB180120207	−1.30 ± 0.04	157 ± 3

Table 3. Spectral parameters of the initial 2 s of long GRBs.

GRB	α	E_{peak} [keV]
GRB081009140	-1.08 ± 0.25	41 ± 4
GRB081215784	-0.46 ± 0.04	699 ± 32
GRB090424592	-1.02 ± 0.05	219 ± 6
GRB090719063	-0.17 ± 0.09	339 ± 14
GRB090804940	-0.45 ± 0.15	114 ± 3
GRB091127976	-1.67 ± 0.07	225 ± 28
GRB100131730	-0.97 ± 0.08	246 ± 15
GRB100324172	0.34 ± 0.09	526 ± 23
GRB100722096	-1.51 ± 0.15	65 ± 7
GRB100829876	-1.00 ± 0.06	247 ± 13
GRB100910818	-0.64 ± 0.87	86 ± 15
GRB101208498	-1.44 ± 0.15	116 ± 10
GRB110817191	-0.54 ± 0.08	275 ± 12
GRB110921912	-1.00 ± 0.09	830 ± 190
GRB111220486	-0.94 ± 0.23	428 ± 149
GRB120129580	-0.95 ± 0.03	347 ± 9
GRB120204054	0.50 ± 0.61	362 ± 73
GRB120217904	-1.16 ± 0.09	322 ± 39
GRB120328268	-1.05 ± 0.23	615 ± 360
GRB120426090	-0.87 ± 0.07	148 ± 3
GRB120728434	-1.32 ± 0.24	204 ± 51
GRB120801920	-0.19 ± 0.50	440 ± 105
GRB130121835	-0.52 ± 0.32	335 ± 67
GRB130228212	-0.75 ± 0.34	262 ± 55
GRB130306991	-0.89 ± 0.90	572 ± 506
GRB130425327	-0.83 ± 0.75	197 ± 66
GRB130502327	-0.69 ± 0.66	165 ± 49
GRB130815660	-1.83 ± 0.31	162 ± 149
GRB130821674	-0.56 ± 0.50	387 ± 156
GRB131108862	-0.68 ± 0.09	460 ± 38
GRB131214705	-0.38 ± 0.16	405 ± 53
GRB131229277	-1.13 ± 0.40	338 ± 202
GRB140213807	-1.32 ± 0.08	279 ± 32
GRB140523129	-0.79 ± 0.06	458 ± 31
GRB140621827	-0.80 ± 0.22	596 ± 233
GRB140801792	-0.29 ± 0.16	128 ± 4
GRB141222298	-1.32 ± 0.33	144 ± 35
GRB150330828	0.09 ± 0.24	358 ± 39
GRB150403913	-0.84 ± 0.24	614 ± 264
GRB150426594	-0.96 ± 0.75	133 ± 39
GRB151227072	-0.88 ± 0.12	181 ± 10
GRB151227218	-1.15 ± 0.12	322 ± 55
GRB151231443	-1.34 ± 0.20	201 ± 26
GRB160113398	-0.58 ± 2.69	96 ± 55
GRB160516237	-1.50 ± 0.45	89 ± 32
GRB160521385	-0.80 ± 0.06	217 ± 6
GRB160724444	-1.13 ± 0.14	475 ± 136
GRB16080225	-0.49 ± 0.03	444 ± 11
GRB160816730	-0.58 ± 0.07	317 ± 16
GRB160910722	0.03 ± 0.80	204 ± 54
GRB161218356	-1.12 ± 0.08	277 ± 22
GRB170207906	1.49 ± 0.80	155 ± 13
GRB170511249	0.05 ± 1.22	96 ± 17
GRB170522657	-0.33 ± 0.13	394 ± 36
GRB170626401	-0.78 ± 0.13	136 ± 5
GRB170802638	-0.44 ± 0.65	316 ± 132
GRB170826819	-0.47 ± 0.17	366 ± 48
GRB171120556	-1.08 ± 0.06	332 ± 23
GRB180120207	-0.74 ± 0.09	210 ± 10

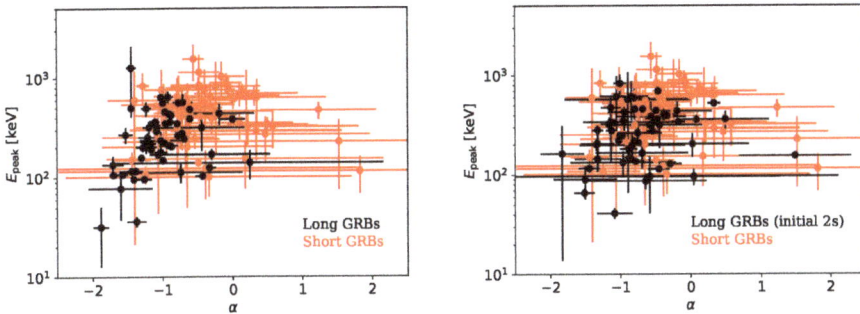

Figure 2. Distribution of E_{peak} versus α between long GRBs and short GRBs (**left**), and the initial 2 s of long GRBs and short GRBs (**right**).

Figure 3 shows a comparison of the histograms of E_{peak} between short GRBs and long GRBs (left panel), and short GRBs and the initial 2 s of long GRBs (right panel). The medians of E_{peak} are 240.5 keV, 479.7 keV and 278.6 keV for long GRBs, short GRBs and the initial 2 s of long GRBs, respectively. The Kolmogorov–Smirnov (K–S) test probabilities of E_{peak} between short GRBs and long GRBs, and short GRBs and the initial 2 s of long GRBs are 2.2×10^{-5} and 5.3×10^{-5}, respectively. Therefore, the K–S test shows that the E_{peak} distributions are drawn from a different population between short GRBs and long GRBs, as well as short GRBs and the initial 2 s of long GRBs.

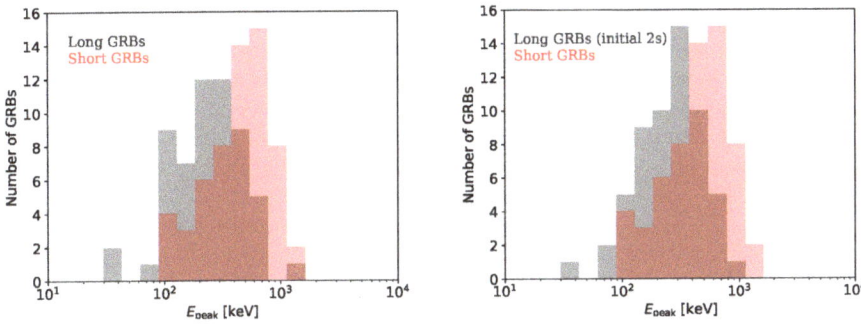

Figure 3. Histograms of E_{peak} between short GRBs and long GRBs (**left**), and short GRBs and the initial 2 s of long GRBs (**right**).

Figure 4 shows a comparison of the histograms of α between short GRBs and long GRBs (left panel), and short GRBs and the initial 2 s of long GRBs (right panel). The medians of α are -1.02, -0.47 and -0.80 for long GRBs, short GRBs and the initial 2 s of long GRBs, respectively. The median of α of the initial 2 s of long GRBs becomes closer to the α distribution of short GRBs. The K–S test probabilities of α between short GRBs and long GRBs, and short GRBs and the initial 2 s of long GRBs are 1.6×10^{-8} and 2.4×10^{-3}, respectively. This statistical test shows that the α distribution of short GRBs becomes closer to that of the initial 2 s of long GRBs.

In summary, the spectra of the initial 2 s of long GRBs show a flatter (harder) α than those of the time-averaged spectra of long GRBs. The distribution of α of the initial 2 s of long GRBs is closer to that of short GRBs. However, the distribution of E_{peak} of the initial 2 s of long GRBs is lower than that of short GRBs, and consistent with the distribution of long GRBs.

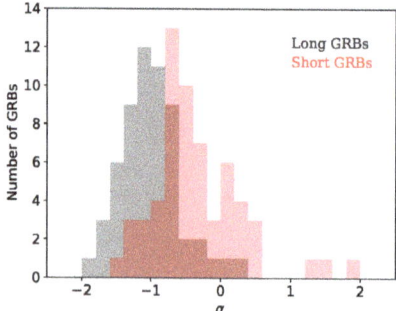
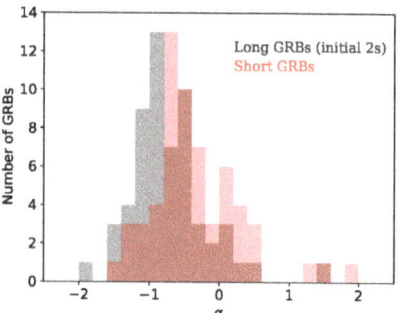

Figure 4. Histograms of α between short GRBs and long GRBs (**left**), and short GRBs and the initial 2 s of long GRBs (**right**).

4. Discussion

Ghirlanda et al. [11] investigated the spectral properties of short and long GRBs using BATSE data. They found that short GRBs have a harder α and a higher E_{peak} compared to those of long GRBs. Furthermore, they also found no difference in α and E_{peak} between the initial 1–2 s of long GRB and short GRB spectra. Our independent analysis shows a systematically harder α for the initial 2 s of long GRBs and their α are closer to those of short GRBs. Although the E_{peak} distributions between short GRBs and the initial 2 s of long GRBs do not show a statistically significant similarity, the E_{peak} distribution of the initial 2 s of long GRBs shows a shift toward a high energy side (Figure 3). Our independent analysis based on the *Fermi* GBM data confirms the findings by Ghirlanda et al. [11].

The radiation process of a prompt GRB emission is described by synchrotron emission via fast cooling electrons. Therefore, there is a strong restriction on the allowed α from $-3/2$ to $-2/3$ e.g., [24]. The limitation of a harder side of α is $-2/3$. The previous study [25] showed that the spectra of 23% of the BATSE GRB samples violated the limit. According to our samples, about 9% of the time-averaged spectra of long GRBs violate the harder side of α by taking into account the error on α. On the other hand, 20% of short GRBs and 24% of the initial 2 s of long GRBs violate this limit. Since the similar fractions of the GRB spectra are violating the synchrotron limit for short GRBs and the initial 2 s of long GRBs, this might indicate that the similar radiation process, which is difficult to achieve by synchrotron emission, is involved in those spectra. Applying the K–S test to the E_{peak} distributions of the spectra violating and non-violating the synchrotron limit for short GRBs, the initial 2 s of the long GRBs and long GRBs, we find K–S test probabilities of 7.6×10^{-3}, 8.2×10^{-2} and 4.4×10^{-1}, respectively.

The initial part of a prompt emission spectrum of long GRBs shows a peculiar characteristic in general. A spectral evolution of a prompt emission follows a hard-to-soft trend e.g., [26]. There is a well-known correlation between intensity and hardness during a burst [27]. However, according to Lu et al. [28], the flux and E_{peak} during a burst do not follow a positive correlation at the initial phase, mainly a rising part of the burst episode. Since a time-averaged spectrum of a long GRB is dominated by the emission from a peak to a tail part of a burst for a single pulse GRB, the distinct spectral characteristic which we see between long GRBs and the initial 2 s of long GRBs is related to the finding of Lu et al. [28]. A detailed study of the initial part of a GRB spectrum, especially the rising part of a GRB emission, will be important to understand the radiation processes of a prompt GRB emission.

Thanks to the gravitational wave detection accompany with a short GRB, the progenitor of a short GRB has been solved. Therefore, it becomes clear that the progenitors of long and short GRBs are different. Although long and short GRBs are originated to a different progenitor, our results suggest that the similar radiation process is involved between those two different classes of GRBs. We suggest that a similar kind of a relativistic jet needs to be launched in both long and short GRBs to explain the similarity in the spectral properties.

Author Contributions: Formal analysis, Y.Y.; Writing—original draft, T.S.; Writing—review and editing, M.S.

Funding: This research received no external funding.

Acknowledgments: We would like to thank the anonymous referees for comments and suggestions that materially improved the paper. This work is supported by MEXT KAKENHI Grant Numbers 17H06357 and 17H06362.

Conflicts of Interest: The authors declare no conflict of interest.

References

1. Abbott, B.P.; Abbott, R.; Abbott, T.D.; Acernese, F.; Ackley, K.; Adams, C.; Adams, T.; Addesso, P.; Adhikari, R.X.; Adya, V.B.; et al. GW170817: Observation of Gravitational Waves from a Binary Neutron Star Inspiral. *Phys. Rev. Lett.* **2017**, *119*, 161101. [CrossRef] [PubMed]
2. Abbott, B.P.; Abbott, R.; Adhikari, R.X.; Ananyeva, A.; Anderson, S.B.; Appert, S.; Arai, K.; Araya, M.C.; Barayoga, J.C.; Barish, B.C.; et al. Multi-messenger Observations of a Binary Neutron Star Merger. *Astrophys. J. Lett.* **2017**, *848*, 59. [CrossRef]
3. Goldstein, A.; Veres, P.; Burns, E.; Briggs, M.S.; Hamburg, R.; Kocevski, D.; Wilson-Hodge, C.A.; Preece, R.D.; Poolakkil, S.; Roberts, O.J.; et al. An Ordinary Short Gamma-ray Burst with Extraordinary Implications: *Fermi* GBM Detection of GRB 170817A. *Astrophys. J. Lett.* **2017**, *848*, 14. [CrossRef]
4. Troja, E.; Piro, L.; Van Eerten, H.; Wollaeger, R.T.; Im, M.; Fox, O.D.; Butler, N.R.; Cenko, S.B.; Sakamoto, T.; Fryer, C.L.; et al. The X-ray Counterpart to the Gravitational-wave Event GW170817. *Nature* **2017**, *551*, 71–74. [CrossRef]
5. Troja, E.; Piro, L.; Ryan, G.; van Eerten, H.; Ricci, R.; Wieringa, M.H.; Lotti, S.; Sakamoto, T.; Cenko, S.B. The Outflow Structure of GW170817 from Late-time Broad-band Observations. *Mon. Not. R. Astron. Soc. Lett.* **2018**, *478*, L18–L23. [CrossRef]
6. Mooley, K.P.; Nakar, E.; Hotokezaka, K.; Hallinan, G.; Corsi, A.; Frail, D.A.; Horesh, A.; Murphy, T.; Lenc, E.; Kaplan, D.L.; et al. A Mildly Relativistic Wide-angle Outflow in the Neutron-star Merger Event GW170817. *Nature* **2018**, *554*, 207–210. [CrossRef] [PubMed]
7. Yamazaki, R.; Ioka, K.; Nakamura, T. X-ray Flashes from Off-Axis Gamma-ray Bursts. *Astrophys. J. Lett.* **2002**, *571*, L31–L35. [CrossRef]
8. Zhang, B.; Dai, X.; Lloyd-Ronning, N.M.; Mészáros, P. Quasi-universal Gaussian Jets: A Unified Picture for Gamma-ray Bursts and X-ray Flashes. *Astrophys. J. Lett.* **2004**, *601*, L119–L122. [CrossRef]
9. Kouveliotou, C.; Meegan, C.A.; Fishman, G.J.; Bhat, N.P.; Briggs, M.S.; Koshut, T.M.; Paciesas, W.S.; Pendleton, G.N. Identification of two classes of gamma-ray bursts. *Astrophys. J.* **1993**, *413*, L101–L104. [CrossRef]
10. Ghirlanda, G.; Ghisellini, G.; Celotti, A. The spectra of short Gamma-ray Bursts. *Astron. Astrophys.* **2004**, *422*, L55–L58. [CrossRef]
11. Ghirlanda, G.; Nava, L.; Ghisellini, G.; Celotti, A.; Firmani, C. Short versus Long gamma-ray bursts: Spectra, energetics, and luminosities. *Astron. Astrophys.* **2009**, *496*, 585–595. [CrossRef]
12. Hakkila, J.; Preece, R.D. Unification of Pulses in Long and Short Gamma-ray Bursts: Evidence from Pulse Properties and their Correlations. *Astrophys. J.* **2011**, *740*, 104. [CrossRef]
13. Ghirlanda, G.; Ghisellini, G.; Nava, L. Short and Long Gamma-ray Bursts: Same Emission Mechanism? *Mon. Not. R. Astron. Soc. Lett.* **2011**, *418*, L109–L113. [CrossRef]
14. Gehrels, N.; Chincarini, G.; Giommi, P.; Mason, K.O.; Nousek, J.A.; Wells, A.A.; White, N.E.; Barthelmy, S.D.; Burrows, D.N.; Cominsky, L.R.; et al. The Swift Gamma-ray Burst Mission. *Astrophys. J.* **2004**, *611*, 1005–1020. [CrossRef]
15. Barthelmy, S.D.; Barbier, L.M.; Cummings, J.R.; Fenimore, E.E.; Gehrels, N.; Hullinger, D.; Krimm, H.A.; Markwardt, C.B.; Palmer, D.M.; Parsons, A.; et al. The Burst Alert Telescope (BAT) on the SWIFT MIDEX Mission. *Space Sci. Rev.* **2005**, *120*, 143–164. [CrossRef]
16. Meegan, C.; Lichti, G.; Bhat, P.N.; Bissaldi, E.; Briggs, M.S.; Connaughton, V.; Diehl, R.; Fishman, G.; Greiner, J.; Hoover, A.S.; et al. The *Fermi* Gamma-ray Burst Monitor. *Astrophys. J.* **2009**, *702*, 791–804. [CrossRef]

17. Lien, A.; Sakamoto, T.; Barthelmy, S.D.; Baumgartner, W.H.; Cannizzo, J.K.; Chen, K.; Collins, N.R.; Cummings, J.R.; Gehrels, N.; Krimm, H.A.; et al. The Third *Swift* Burst Alert Telescope Gamma-ray Burst Catalog. *Astrophys. J.* **2016**, *829*, 47. [CrossRef]
18. Sakamoto, T.; Barthelmy, S.D.; Barbier, L.; Cummings, J.R.; Fenimore, E.E.; Gehrels, N.; Hullinger, D.; Krimm, H.A.; Markwardt, C.B.; Palmer, D.M.; et al. The First *Swift* BAT Gamma-ray Burst Catalog. *Astrophys. J. Suppl. Ser.* **2008**, *175*, 179–190. [CrossRef]
19. Sakamoto, T.; Barthelmy, S.D.; Baumgartner, W.H.; Cummings, J.R.; Fenimore, E.E.; Gehrels, N.; Krimm, H.A.; Markwardt, C.B.; Palmer, D.M.; Parsons, A.M.; et al. The Second *Swift* Burst Alert Telescope Gamma-ray Burst Catalog. *Astrophys. J. Suppl. Ser.* **2011**, *195*, 27. [CrossRef]
20. Sakamoto, T.; Pal'Shin, V.; Yamaoka, K.; Ohno, M.; Sato, G.; Aptekar, R.; Barthelmy, S.D.; Baumgartner, W.H.; Cummings, J.R.; Fenimore, E.E.; et al. Spectral Cross-Calibration of the Konus–Wind, the Suzaku/WAM, and the Swift/BAT Data Using Gamma-ray Bursts. *Publ. Astron. Soc. Jpn.* **2011**, *63*, 215–277. [CrossRef]
21. Gruber, D.; Goldstein, A.; von Ahlefeld, V.W.; Bhat, P.N.; Bissaldi, E.; Briggs, M.S.; Byrne, D.; Cleveland, W.H.; Connaughton, V.; Diehl, R.; et al. The *Fermi* GBM Gamma-ray Burst Spectral Catalog: Four Years of Data. *Astrophys. J. Suppl. Ser.* **2014**, *211*, 27. [CrossRef]
22. Von Kienlin, A.; Meegan, C.A.; Paciesas, W.S.; Bhat, P.N.; Bissaldi, E.; Briggs, M.S.; Burgess, J.M.; Byrne, D.; Chaplin, V.; Cleveland, W.; et al. The Second *Fermi* GBM Gamma-ray Burst Catalog: The First Four Years. *Astrophys. J. Suppl. Ser.* **2014**, *211*, 13. [CrossRef]
23. Bhat, P.N.; Meegan, C.A.; von Kienlin, A.; Paciesas, W.S.; Briggs, M.S.; Burgess, J.M.; Burns, E.; Chaplin, V.; Cleveland, W.H.; Collazzi, A.C.; et al. The Third *Fermi* GBM Gamma-ray Burst Catalog: The First Six Years. *Astrophys. J. Suppl. Ser.* **2016**, *223*, 18. [CrossRef]
24. Sari, R.; Piran, T.; Narayan, R. Spectra and Light curves of Gamma-ray Bursts Afterglows. *Astrophys. J. Lett.* **1998**, *497*, L17–L20. [CrossRef]
25. Preece, R.D.; Briggs, M.S.; Mallozzi, R.S.; Pendleton, G.N.; Paciesas, W.S.; Band, D.L The Synchrotron Shock Model Confronts a "Line of Death" in the BATSE Gamma-ray Burst Data. *Astrophys. J. Lett.* **1998**, *506*, L23–L26. [CrossRef]
26. Norris, J.P.; Share, G.H.; Messina, D.C.; Dennis, B.R.; Desai, U.D.; Cline, T.L.; Matz, S.M.; Chupp, E.L. Spectral Evolution of Pulse Structures in Gamma-ray Burst. *Astrophys. J.* **1986**, *301*, 213–219. [CrossRef]
27. Golenetskii, S.V.; Mazets, E.P.; Aptekar, R.L.; Ilinskii, V.N. Correlation between Luminosity and Temperature in Gamma-ray Burst Sources. *Nature* **1983**, *306*, 451–453. [CrossRef]
28. Lu, R.J.; Hou, S.J. Liang, E.W. The E_{peak}-Flux Correlation in the Rising and Decaying Phases of Gamma-ray Burst Pulses: Evidence for Viewing Angle Effect? *Astrophys. J.* **2010**, *720*, 1146–1154. [CrossRef]

© 2018 by the authors. Licensee MDPI, Basel, Switzerland. This article is an open access article distributed under the terms and conditions of the Creative Commons Attribution (CC BY) license (http://creativecommons.org/licenses/by/4.0/).

Article

Reverse Shock Emission from Short GRBs

Nicole Lloyd-Ronning [1,2]

1. Center for Theoretical Astrophysics, Los Alamos National Lab, Los Alamos, NM 87544, USA; lloyd-ronning@lanl.gov
2. Department of Math and Science, University of New Mexico, Los Alamos, NM 87544, USA

Received: 14 August 2018; Accepted: 25 September 2018; Published: 28 September 2018

Abstract: We investigate the expected radio emission from the reverse shock of short GRBs, using the fitted afterglow parameters. In light of recent results suggesting that in some cases the radio afterglow is due to emission from the reverse shock, we examine the extent to which this component is detectable for short GRBs. In some GRBs, the standard synchrotron shock model predicts detectable radio emission from the reverse shock when none was seen. Many physical parameters play a role in these estimates, and our results highlight the need to explore the fundamental processes involved in GRB particle acceleration and emission more deeply. However, with a more rapid follow-up, we can test our standard model of GRBs, which predicts an early, radio bright reverse shock in many cases.

Keywords: gamma-ray bursts

1. Introduction

Perhaps the most robust model for short gamma-ray bursts (SGRBs) is the merger of two compact objects, such as two neutrons stars (NS-NS) or a neutron star and a black hole (NS-BH). The timescales and energetics involved in the merger have always made this a plausible model for SGRBs [1,2], but other clues including the location of these bursts in their host galaxies, the lack of associated supernovae and of course the recent detection of gravitational waves from a neutron star merger coincident with a SGRB [3] have provided convincing evidence that these bursts are associated with the older stellar populations expected of compact objects [4–12].

There has been a concerted effort to follow up on short GRBs with the goal of detecting the afterglow and potentially learning more about this class of gamma-ray bursts (for a review, see [13]). To date, about 93%, 84% and 58% of SGRBs have been followed up in the X-ray, optical and radio spectra, respectively [14]. Of these follow-up efforts, 74% have an X-ray afterglow, 34% have been seen in the optical and only 7% detected in the radio.

Recently, Lloyd-Ronning and Fryer investigated a sample of long GRBs that were followed up in the radio and found that bright bursts (with isotropic equivalent energy $E_{iso} > 10^{52} erg$) without radio afterglows had a significantly shorter intrinsic prompt duration. They explored various reasons for the lack of afterglow in the context of different progenitor models; one possibility for the lack of radio afterglow is that this emission comes primarily from the reverse shock and that those with no radio afterglow are in a parameter space with a weak reverse shock signal.

On the other hand, Laskar et al. [16,17] and Alexander et al. [18] have recently reported the detection of a distinct reverse shock component in the afterglows of GRB130427A, GRB160509A and 160625B. They suggested that the external medium density must be low ($n < 1$ cm^{-3}) in order to give a long-lived radio afterglow from the reverse shock (the low density allows for the emitting electrons to be in the so-called slow-cooling regime, thereby giving rise to longer-lived reverse shock emission). These results combined with those from Lloyd-Ronning and Fryer prompted us to investigate why more short GRBs (with their presumed low circumstellar densities) do not have a detected radio afterglow from the reverse shock.

Using the multi-band afterglow fits from Fong et al. [14], we explore the detectability of the reverse shock component from SGRBs. Using their fitted parameters for emission from the forward shock, we estimate the emission from the reverse shock, using the same formalism as [16,17]. We find that in some cases (depending on the microphysical parameters), there should be a detectable radio signal at the time of the afterglow follow-up, when none was seen.

Our paper is organized as follows. In Section 2, we describe how we calculate the radio flux from the forward and reverse shock using the standard formalism of synchrotron emission from a relativistic jet, using the fitted parameters from [14]. In Section 3, we present our results. We find that most of the reverse shock emission occurs too early to be detected in the radio, but in some cases, this emission should have been detected. In Section 4, we summarize and present our conclusions.

2. Materials and Methods

Fong et al. [14] carried out an extensive effort, compiling all of the available afterglow data for 103 SGRBs and fitting these data to the standard synchrotron forward external shock model. Table 3 from Fong et al. [14] gives the results of these fits, in particular the values of p, ϵ_B, the average isotropic kinetic energy E_{iso} and the external density n (assumed a constant, as expected for NS-NS or NS-BH progenitors). Note that they assume the fraction of energy in the electrons is a fixed value of $\epsilon_e = 0.1$. They performed two sets of fits to each burst: one in which the fraction of energy in the magnetic field ϵ_B is 0.1, and one in which the value of $\epsilon_B = 0.01$. If neither gave an acceptable fit, they allowed ϵ_B to be a free parameter (hence explaining the couple of entries with $\epsilon_B \neq 0.1$ or 0.01).

We point out that four individual bursts were detected by Fong et al. [14] in the radio band. These bursts are GRB050724A, GRB051221A, GRB130603B and GRB140903A. Table 1 of this paper gives the time of observation in days and the flux in μJy detected at these times for these SGRBs.

Table 1. Radio afterglow detections of short GRBs.

GRB	t_{obs} (Days)	Flux (μJy)
150724A	0.57, 1.68	173, 465
051221A	0.91	155
130603B	0.37, 1.43	125, 65
140903A	0.4, 2.4, 9.2	110, 187, 81

In the standard picture of a relativistic external blast wave, the onset of the afterglow occurs around the deceleration time t_{dec}, i.e., when the blast wave has swept up enough external material to begin to decelerate $t_{dec} \propto (E/n)^{1/3} \Gamma^{-8/3}$ [19], where E is the energy in the blast wave, n is the external particle number density and Γ is the Lorentz factor of the blast wave. One can calculate the characteristic synchrotron break frequencies at this time, depending on the global and microphysical parameters of the burst. These expressions are given in Table 2 of [20] for both a constant density and wind medium. Figure 1 shows the characteristic break frequencies (and the corresponding flux at these frequencies), using the parameters fitted from the data of Fong et al. [14] at the deceleration time (when the afterglow begins). The light blue dots indicate the self-absorption frequency ν_a of the forward shock, the green dots the frequency corresponding to the minimum or characteristic electron energy ν_m of the forward shock and the pink dots the so-called cooling frequency ν_c of the forward shock (see, e.g., Sari, Piran, & Narayan [21] for more detailed explanations of these frequencies). In general, $\nu_a < \nu_m < \nu_c$ for the forward shock component. The red stars indicate the minimum electron frequency for the reverse shock, $\nu_{m,RS} \approx \nu_m/\Gamma^2$ (note that this assumes the fraction of energy in the magnetic field is roughly the same for the forward and reverse shock, as explained below). Again, to calculate both the characteristic frequencies and the fluxes at these frequencies, we employed the expression given in Table 2 of [20].

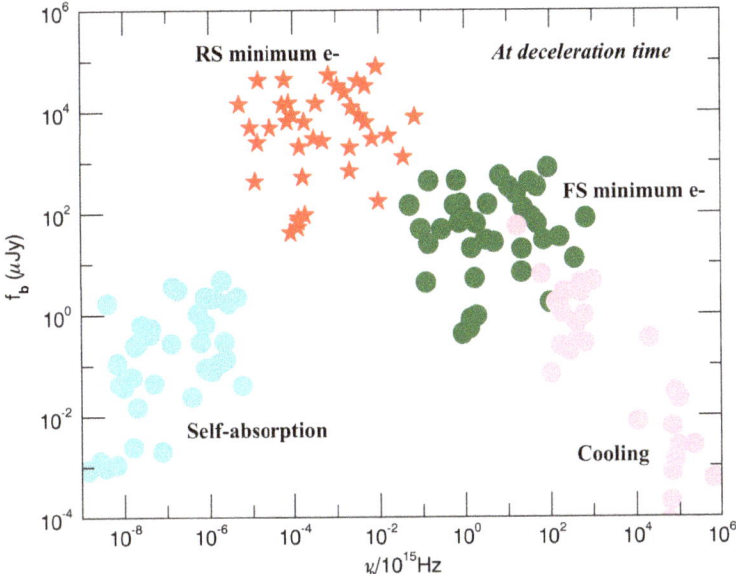

Figure 1. Flux at the characteristic frequency vs. characteristic frequency ν_b (normalized to 10^{15} Hz from a synchrotron spectrum in a standard external shock model, using data from Fong et al. [14], in which $\epsilon_B = 0.01$ was employed in their fits. The light blue dots indicate the forward shock self-absorption frequency ν_a, the green dots the frequency corresponding to the minimum or characteristic electron energy ν_m and the pink dots the so-called cooling frequency ν_c. The red stars indicate the minimum or characteristic electron frequency for the reverse shock, $\nu_{m,RS}$.

2.1. Jet Reverse Shock

There have been many studies of the reverse shock from a relativistic blast wave (e.g., Meszaros & Rees [22], Sari & Piran [23], Kobayashi [24], Zhang et al. [25], Kobayashi & Zhang [26], Zou et al. [27] and the references therein), and the early-time radio flare observation of GRB 990123 has been attributed to the reverse shock [28,29]. In addition, Soderberg & Ramirez-Ruiz [30] examined the expected strength of the reverse shock in six long GRBs and were able to constrain the hydrodynamic evolution and bulk Lorentz factors of these bursts from this component.

As pointed out by these references and others, the evolution of the flux and break frequencies in the reverse shock depends on whether the blast wave is Newtonian or relativistic (among other factors), which in turn is related to the shell thickness Δ estimated from the observed duration T by $\Delta \sim cT/(1+z)$. For a thick shell, $\Delta > l/2\Gamma^{8/3}$, where l is the Sedov length in an interstellar medium $\equiv (3E/4\pi n m_p c^2)^{1/3}$, the reverse shock has time to become relativistic and the standard Blandford–McKee solution applies. For a thin shell, the reverse shock remains Newtonian, and the Lorentz factor of this shock evolves as $\Gamma_{RS} \sim r^{-g}$, with $g \sim 2$ [24]. Short bursts with $T < 1$ s are likely in the thin shell—and therefore Newtonian—regime. However, we note that for a range of g values, the time evolution of the flux and characteristic frequencies is fairly similar between the relativistic and Newtonian regimes.

This standard treatment overly simplifies reverse shock emission by separating it into two distinct regimes (thick shell and thin shell), when in reality, the shell thickness covers a range of values and could fall in between these regimes [31]. This simplified treatment also assumes that ϵ_B and ϵ_e are constant in the shell, which is not necessarily justified. An evolving ϵ_B and ϵ_e will complicate the evolution of the flux and characteristic frequencies and allow an additional degree of freedom in the treatment of the reverse shock.

However, generally speaking, because of the higher mass density in the shell, the peak flux in the reverse shock $f_{p,RS}$ will be higher by a factor of Γ relative to the forward shock,

$$f_{p,RS} \approx \Gamma f_{p,FS} \tag{1}$$

but the minimum electron frequency in the reverse shock $\nu_{m,RS}$ will be lower by a factor of Γ^2,

$$\nu_{m,RS} \approx \nu_{m,FS}/\Gamma^2 \tag{2}$$

assuming the forward and reverse shock have the same fraction of energy in the magnetic field (also not necessarily a well-justified assumption; see the discussion below). For the purposes of comparing with others' analyses of reverse shock emission [16,17], we employ this prescription in our estimates below.

2.1.1. Self-Absorbed Reverse Shock

Because we are examining the radio emission, we need to be concerned with synchrotron self-absorption; under certain conditions, lower energy photons are self-absorbed, and the flux is suppressed. Self-absorption may be particularly relevant in the region of reverse shock, where the density is higher relative to the forward shock region. Resmi & Zhang [32] calculated the relevance of the self-absorption frequency and flux in the reverse shock, before and after shock crossing. For our purposes—because we are looking at later time radio emission—we consider the frequencies and fluxes after the shock crosses the thin shell (but, see their Appendix A.1 for expressions in all ranges of parameter space).

Roughly, at the high radio frequencies we are considering here, the flux at the time of the peak can be obtained from Equation 30 of Resmi & Zhang [32]:

$$f_{p,RS} = f_{p,RS,\nu_m}(\nu_{a,RS}/\nu_{m,RS})^{-\beta} \tag{3}$$

where $\beta = (p-1)/2$.

The reverse shock flux is suppressed at a minimum by factors ranging from about 0.3–0.01. We emphasize again, therefore, that our estimates are upper limits to the emission from the reverse shock.

3. Results

Figure 2 shows our estimates of the peak flux at ν_m for the forward (blue circles) and reverse (red stars) shocks at the time ν_m reaches the radio band of 8.46 GHz. The left panels are for a Lorentz factor $\Gamma = 100$, while the right panels are for $\Gamma = 10$. The top panels of Figure 2 utilize the $\epsilon_B = 0.1$ fits from Fong et al. [14], while the bottom panels utilize the parameters from their $\epsilon_B = 0.01$ fits. Note that Fong et al. [14] reported the median of the observing time response for the radio afterglow follow-up observations tobe about 1 day. This is reflected in Figure 2 by the vertical shaded regions. The horizontal shaded regions show roughly the detector flux limits. The red dashed lines show the standard synchrotron flux decay as a function of time for a few representative bursts, assuming the reverse shock has become relativistic and a Blandford–McKee solution applies. This temporal decay was computed using the fitted parameters of [14] (which determine the relative values of the characteristic synchrotron break frequencies) and the expressions for synchrotron flux given in Table 2 of [20].

We point out that although many sGRBs have apparent non-thermal gamma-ray photons that constrain the Lorentz factor to be large, $\Gamma \geq 100$ (a compact region will be optically thick to pair production at gamma-ray energies, unless the region is moving relativistically, [33]), some sGRBs do not impose such stringent constraints, and a $\Gamma \sim 10$ is sufficient to allow for their non-thermal spectra (the most famous example is GRB170817 [3], but see also [34] which show a sample of GRBs with a lack of high energy photons). We display both $\Gamma = 100$ and $\Gamma = 10$ not necessarily to argue for low Lorentz factor sGRBs, but to show how the reverse shock flux and its peak time vary as a function of the Lorentz factor.

It is clear that in the context of this model, many of the reverse shock bursts were missed because they tended to peak before the beginning of the observations. The red dashed lines show the flux decay as a function of time for a few representative bursts. Several bursts' flux values crossed distinctly through the observational window and above the flux limit as they faded.

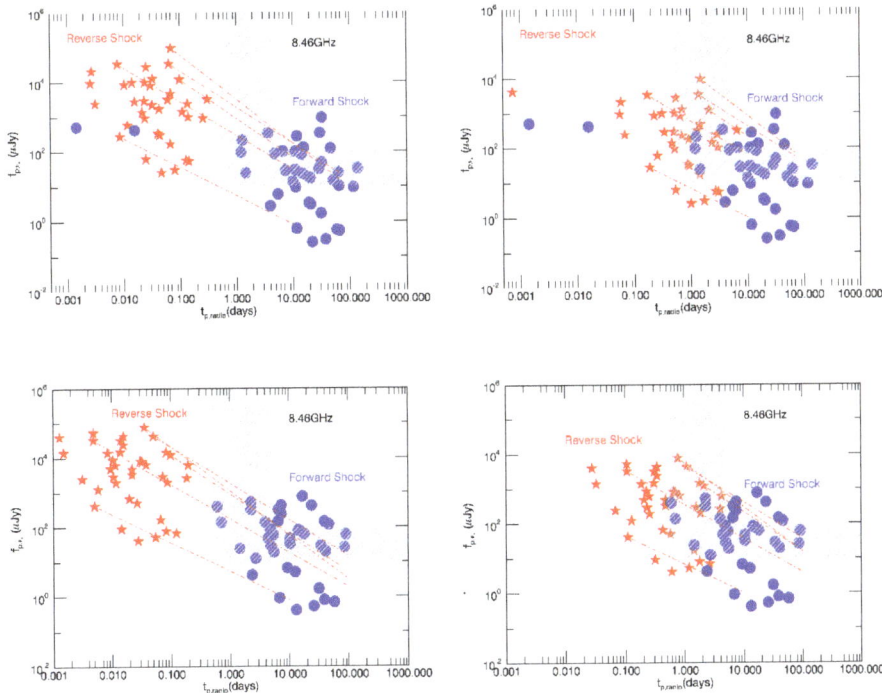

Figure 2. Estimates of the peak flux from the forward (blue dots) and reverse (red stars) shock from synchrotron emission. The vertical shaded region marks the temporal window when radio follow-up observations began for this sample. The horizontal shaded region marks the rough observational flux density limit. Top left panel: $\epsilon_B = 0.1, \epsilon_e = \epsilon_B, \Gamma = 100$. Top right panel: $\epsilon_B = 0.1, \epsilon_e = \epsilon_B, \Gamma = 10$. The red dashed lines show the flux decay as a function of time for a few representative bursts. Bottom left panel: $\epsilon_B = 0.01, \epsilon_e = 0.1, \Gamma = 100$. Bottom right panel: $\epsilon_B = 0.01, \epsilon_e = 0.1, \Gamma = 10$. The red dashed lines show the predicted flux decay as a function of time for a few representative bursts.

Note that a few bursts that went undetected in the radio (i.e., not one of the four listed in Table 1) indicated a forward shock component in the observational window in Figure 2. However, on closer examination, comparing the time of observations of these particular bursts to the predicted time of the peak (at ν_m), we see that the radio observations occurred well before the predicted peak time (which occurred at $\approx$ 10's of days in all of our models), and may be why it was not detected. However, as discussed above, the reverse shock emission fell above the flux limit for several bursts (in particular for GRB11112A, GRB121226A, GRB131004A, GRB150101B) during the time of their radio observations, particularly for the lower Lorentz factor cases ($\Gamma = 10$; right panels of Figure 2). The fact that this emission was not detected suggests that, at least in some cases, the reverse shock flux derived from this standard prescription of GRB afterglow emission was overly simplistic and gave misleading values for the flux (we again emphasize that we are looking in the optically thin limit here, and it may also be that the reverse shock emission was self-absorbed in these cases).

In any case, it is clear that rapid follow-up in the radio gave us a better chance to detect and/or constrain this important component, potentially breaking some of the degeneracies amongst the physical parameters in the models and allowing us to better understand the physics behind SGRB emission.

4. Conclusions

We have investigated radio emission from short gamma-ray bursts, using fits from existing broadband afterglow data [14] in the context of the standard synchrotron shock model for GRB emission. In particular, we have looked at the peak flux from the forward and reverse shock components of the relativistic jet. We find in some cases that the reverse shock component should have been detected in the context of this standard model. The lack of detection suggests any number of oversimplifications in the model, including potentially variable microphysical parameters, a misestimated bulk Lorentz factor and/or not properly accounting for self-absorption.

We can get additional important information on short gamma-ray bursts if there is rapid follow-up (<1 day) in the radio; this will give the best chance of detecting the reverse shock emission component. High Lorentz factor outflows the $\Gamma \sim 100$ peak very early ($t \sim 0.05$ day) and may be very challenging to detect. The lower Lorentz factor outflows the $\Gamma \sim 10$ peak later and gives us a better chance of temporally catching the reverse shock; however, the flux will be lower for the less relativistic outflows. The circumburst density must also be low enough to allow for a slow-cooling reverse shock (as mentioned in Laskar et al. [16,17]), but such densities are expected for compact object binary progenitors of sGRBs.

The electromagnetic signals can be very sensitive to the values of the microphysical parameters, such as the fraction of energy in the electrons and magnetic field, so a concerted effort to more definitively constrain those parameters from a theoretical standpoint would be helpful in breaking the degeneracies and pinning down global burst parameters like kinetic energy, circumburst density, etc. These latter parameters can help constrain the progenitor.

Once again, more rapid follow-up (ideally within hours) with greater sensitivity (from 10 µJy to $\approx$100 µJy) could produce a significant number of detections of radio emission from the reverse shock. A lack of detection would also constrain models to some extent and point us toward areas in which we are oversimplifying our treatment of GRB emission.

We point out again, however, that the radio emission is just one piece of the puzzle in understanding GRB emission, and it is only through multi-wavelength follow-up that we will really be able to constrain the underlying physics of the outflow producing gamma-ray bursts. Efforts in this vein are particularly timely in light of the near era of gravitational wave detection from a double neutron star merger. A better understanding of the various components of electromagnetic emission from these objects will provide a more complete picture of these systems and ultimately help us understand their role in the context of stellar evolution in the universe.

Funding: This research received no external funding.

Acknowledgments: Work at LANLwas done under the auspices of the National Nuclear Security Administration of the U.S. Department of Energy at Los Alamos National Laboratory LA-UR-17-24900.

Conflicts of Interest: The author declares no conflict of interest.

References

1. Eichler, D.; Livio, M.; Piran, T.; Schramm, D.N. A Spectacular Radio Flare from XRF 050416a at 40 Days and Implications for the Nature of X-ray Flashes. *Nature* **1989**, *240*, 126. [CrossRef]
2. Narayan, R.; Paczynski, B.; Piran, T. Using the Active Collimator and Shield Assembly of an EXIST-Type Mission as a Gamma-Ray Burst Spectrometer. *Astrophys. J.* **1992**, *395*, L83 [CrossRef]
3. Abbott, B.P.; Abbott, R.; Abbott, T.D.; Acernese, F.; Ackley, K.; Adams, C.; Adams, T.; Addesso, P.; Adhikari, R.X.; Adya, V.B.; et al. GW170817: Observationso of Gravitational Waves from a Binary Neutron Star Inspiral. *Phys. Rev. Lett.* **2017**, *119*, 1101. [CrossRef] [PubMed]
4. Fox, D.B.; Frail, D.A.; Hurley, J.C. The afterglow of GRB 050709 and the nature of the short-hard gamma-ray bursts. *Nature* **2005**, *437*, 845–850. [CrossRef] [PubMed]
5. Soderberg, A.; Nakar, E.; Berger, E.; Kulkarni, S.R. Late-time radio observations of 68 type Ibc supernovae: Strong constraints on off-axis gamma-ray bursts. *Astrophys. J.* **2006**, *638*, 930. [CrossRef]
6. Berger, E.; Cenko, S.B.; Fox, D.B.; Cucchiara, A. Discovery of the very red near-infrared and optical afterglow of the short-duration GRB 070724A. *Astrophys. J.* **2009**, *704*, 877. [CrossRef]
7. Kocevski, D.; Thöne, C.C.; Ramirez-Ruiz, E.; Bloom, J.S.; Granot, J.; Butler, N.R.; Perley, D.A.; Modjaz, M.; Lee, W.H.; Cobb, B.E. Limits on radioactive powered emission associated with a short-hard GRB 070724A in a star-forming galaxy. *Mon. Not. R. Astron. Soc.* **2010**, *404*, 963–974. [CrossRef]
8. Leibler, C.N.; Berger, E. The stellar ages and masses of short gamma-ray burst host galaxies: Investigating the progenitor delay time distribution and the role of mass and star formation in the short gamma-ray burst rate. *Astrophys. J.* **2010**, *725*, 1202. [CrossRef]
9. Fong, W.; Berger, E.; Fox, D.B. Hubble space telescope observations of short gamma-ray burst host galaxies: Morphologies, offsets, and local environments. *Astrophys. J.* **2010**, *708*, 9. [CrossRef]
10. Berger, E. A short gamma-ray burst "No-host" problem? Investigating large progenitor offsets for short GRBs with optical afterglows. *Astrophys. J.* **2010**, *722*, 1946. [CrossRef]
11. Fong, W.; Berger, E.; Chornock, R.; Margutti, R.; Levan, A.J.; Tanvir, N.R.; Tunnicliffe, R.L.; Czekala, I.; Fox, D.B.; Perley, D.A. Demographics of the galaxies hosting short-duration gamma-ray bursts. *Astrophys. J.* **2013**, *769*, 56. [CrossRef]
12. Fong, W.; Berger, E.; Metzger, B.D.; Margutti, R.; Chornock, R.; Migliori, G.; Foley, R.J.; Zauderer, B.A.; Lunnan, R.; Laskar, T. Short GRB 130603B: Discovery of a jet break in the optical and radio afterglows, and a mysterious late-time X-ray excess. *Astrophys. J.* **2014**, *780*, 118. [CrossRef]
13. Berger, E. Short-duration gamma-ray bursts. *Annu. Rew.* **2014**, *52*, 43–105. [CrossRef]
14. Fong, W.; Berger, E.; Margutti, R.; Zauderer B.A. A decade of short-duration gamma-ray burst broadband afterglows: Energetics, circumburst densities, and jet opening angles. *Astrophys. J.* **2015**, *815*, 102. [CrossRef]
15. Lloyd-Ronning, N.M.; Fryer, C.L. On the lack of a radio afterglow from some gamma-ray bursts–insight into their progenitors? *Mon. Not. R. Astron. Soc.* **2017**, *467*, 3413–3423. [CrossRef]
16. Laskar, T.; Berger, E.; Zauderer, B.A.; Margutti, R.; Soderberg, A.M.; Chakraborti, S.; Lunnan, R.; Chornock, R.; Chandra, P.; Ray, A. A reverse shock in GRB 130427A. *Astrophys. J.* **2013**, *776*, 119. [CrossRef]
17. Laskar, T.; Alexander, K.D.; Berger, E.; Fong, We.; Margutti, R.; Shivvers, I.; Williams, P.K.G.; Kopač, D.; Kobayashi, S.; Mundell, C. A Reverse Shock in Grb 160509a. *Astrophys. J.* **2016**, *833*, 88. [CrossRef]
18. Alexander, K.D. Radio Emission from Short Gamma-ray Bursts in the Multi-Messenger Era. *arXiv* **2017**, arXiv:1705.08455.
19. Blandford, R.D.; McKee, C.F. Fluid dynamics of relativistic blast waves. *Phys. Fluids* **1976**, *19*, 1130. [CrossRef]
20. Granot, J.; Sari, R. The shape of spectral breaks in gamma-ray burst afterglows. *Astrophys. J.* **2002**, *568*, 820. [CrossRef]
21. Sari, R.; Piran, T.; Narayan, R. Spectra and light curves of gamma-ray burst afterglows. *Astrophys. J.* **1998**, *497*, L17. [CrossRef]
22. Meszaros, P.; Rees, M.J. Optical and Long-wavelength Afterglow from Gamma-ray Bursts. *Astrophys. J.* **1997**, *476*, 231. [CrossRef]
23. Sari, R.; Piran, T. Predictions for the very early afterglow and the optical flash. *Astrophys. J.* **1999**, *520*, 641. [CrossRef]
24. Kobayashi, S. Light curves of gamma-ray burst optical flashes. *Astrophys. J.* **2000**, *545*, 807. [CrossRef]

25. Zhang, B.; Kobayashi, S.; Meszaros, P. Gamma-ray burst early optical afterglows: Implications for the initial Lorentz factor and the central engine. *Astrophys. J.* **2003**, *595*, 950. [CrossRef]
26. Kobayashi, S.; Zhang, B. Early optical afterglows from wind-type gamma-ray bursts. *Astrophys. J.* **2003**, *597*, 455. [CrossRef]
27. Zou, Y.C.; Wu, X.F.; Dai, Z.G. Early afterglows in wind environments revisited. *Mon. Not. R. Astron. Soc.* **2005**, *363*, 93–106. [CrossRef]
28. Kulkarni, S.; Frail, D.A.; Sari, R. Discovery of a radio flare from GRB 990123. *Astrophys. J.* **1999**, *522*, L97. [CrossRef]
29. Nakar, E.; Piran, T. GRB 990123 revisited: Further evidence of a reverse shock. *Astrophys. J.* **2005**, *619*, L147. [CrossRef]
30. Soderberg, A.; Ramirez-Ruiz, E. Flaring up: Radio diagnostics of the kinematic, hydrodynamic and environmental properties of gamma-ray bursts. *Mon. Not. R. Astron. Soc.* **2003**, *345*, 854–864. [CrossRef]
31. Kopac, D. Radio flares from gamma-ray bursts. *Astrophys. J.* **2015**, *806*, 179. [CrossRef]
32. Resmi, L.; Zhang, B. Gamma-ray burst reverse shock emission in early radio afterglows. *Astrophys. J.* **2016**, *825*, 48. [CrossRef]
33. Lithwick, Y.; Sari, R. Lower limits on Lorentz factors in gamma-ray bursts. *Astrophys. J.* **2001**, *555*, 540L. [CrossRef]
34. Burgess, J.M.; Greiner, J.; Bégué, D.; Berlato, F. A Bayesian Fermi-GBM Short GRB Spectral Catalog. *arXiv* **2017**, arXiv:1710.08362.

© 2018 by the author. Licensee MDPI, Basel, Switzerland. This article is an open access article distributed under the terms and conditions of the Creative Commons Attribution (CC BY) license (http://creativecommons.org/licenses/by/4.0/).

MDPI
St. Alban-Anlage 66
4052 Basel
Switzerland
Tel. +41 61 683 77 34
Fax +41 61 302 89 18
www.mdpi.com

Galaxies Editorial Office
E-mail: galaxies@mdpi.com
www.mdpi.com/journal/galaxies

www.ingramcontent.com/pod-product-compliance
Lightning Source LLC
LaVergne TN
LVHW071444100526
838202LV00088B/6809